For Lilla and Bill
Thanks for everything

To know the future one must understand the past
Niccolò Machiavelli

Forecasts, Famines and Freezes

FORECASTS, FAMINES AND FREEZES

Climate and Man's Future

John Gribbin

WALKER AND COMPANY
New York

First published in the United States of America in 1976 by
the Walker Publishing Company, Inc.

ISBN: 0-8027-0483-2

Library of Congress Catalog Card Number: 74-31920

Printed in the United States of America.

10 9 8 7 6 5 4 3 2

Contents

Forecasts, Famines and Freezes

Introduction

Early in 1973 through my work on the science journal *Nature* I became aware of a growing number of papers being published in the scientific 'literature', and being submitted to *Nature,* which related in one way or another to the theme of climatic change. By the middle of that year, when the news of the continuing droughts in the Sahel region of Africa, in India and in Ethopia broke, the trickle of climatic change publications had become something of a small flood – but there was still no apprehension that the climate of the whole world might be changing, and that all of us might be affected, not just the unfortunate inhabitants of the tropical zones where disaster situations had already arrived.

One reason for this was the wide diversity of scientists looking at the various aspects of the problem. Some astronomers looked at how the Sun's activity affects the upper atmosphere; some meteorologists looked at how the monsoon pattern over India had changed; some climatologists studied historical trends of climate; and so on. Nobody, it seemed, was putting the bits of the jigsaw together to provide a clear picture which could be understood by the non-specialist – let alone the non-scientist.

So it was a deliberate policy of mine by mid-1973 to report all the different developments relating to climatic change, either in *Nature* or the Science Report of *The Times,* or in other journals such as *New Scientist,* and *Environment and Change,* for which I was writing. The message obviously got through, for in July of 1974 I was awarded a Travelling Fellowship by Glaxo for opening up this area of study to regular reporting by science journalists. It's always nice to get an award for your work – but my greatest pleasure in the award came from the realization that climatic change was at last being recognized as probably the most important single area of scientific research (in terms of the potential effect on human lives) of

Introduction
modern times.

The award also meant that I had the opportunity to travel in the U.S. and Canada to visit some of the people who are tackling the problem of climatic change and of predicting future climatic trends. So in half of this book I have put together a lot of the evidence which has accumulated to show how the climate is changing, and why. Some of this is completely new; a large part draws heavily on pieces I have written as individual news stories but which have not been gathered together to make a coherent story before. And in the rest I have described some of the work being done at the research centres that I visited in the autumn of 1974 using the travel award from Glaxo.

I am deeply grateful to Glaxo for providing this opportunity to look further into the question of climatic change, and to my then employers at Macmillan Journals both for their original forbearance in allowing me to ride for so long what must have originally seemed to them something of a hobby horse and for providing the time for me to travel and carry out the investigation reported later in the book. The authors and publishers mentioned in figure captions kindly gave permission for their figures to be reproduced here.

1 The Climatic Threat

'Disaster drought threat to prices', 'Drought may send beef soaring', 'Farmers must adapt to drier Britain'. These are all headlines from British national papers, and all appeared during the summer of 1974, when the third successive dry spring and early summer at last opened people's eyes to the reality of the climatic change we are now experiencing. Stories of drought in Africa, Ethopia and India had already received publicity and gained public sympathy – but the situation became much more urgent for the industrialized nations of the world when it was realized that the same climatic changes that are now affecting the tropics are having an influence on agriculture and industry at higher latitudes.

These headlines, and others like them, left me in no doubt about the use to which I should put the research funds which came my way in the Fall of 1974 – to find out more about the global extent of the problem of climatic change, and to report on the problem to others who may have been worried by such headlines. I soon found that the extent of the problem was matched only by the extent of the general concern that something has gone wrong with the weather; in one issue of the *New York Times* which I came across while writing this book (May 21st, 1975) two separate stories puzzled over the climatic threat under the headlines 'Sun cycle indicates severe drought in '76' and 'Scientists ask why world climate is changing'. You can find out more about that Sun cycle effect in Chapter 5, and the second headline just about sums up this whole book. But first, just how urgent are these climatic problems? If they were just local in scope, then any hard hit country could rely on aid from abroad; but the example of how the British climatic problems are interlocked with those of the U.S. and the rest of the world – the situation which attracted my attention in 1974 – shows that no country today is sufficiently free from climatic problems to be able to offer much help to the hardest hit.

The climatic threat made itself felt first in its effect on agriculture and food prices, but it is also a very real threat to industry. Declining rainfall is enough of a problem in itself; but according to many climatologists the world is also in a phase of declining temperatures, so that, among other factors, there must inevitably be an increase in demand for fuel for heating – and all this at a time when we are in the midst of an energy crisis!

So these are urgent problems which should be tackled as matters of priority by the scientifically knowledgeable nations of the world. Already a great deal is known about how changes in the circulation pattern of the Earth's atmosphere have changed the global climate, and there are hints of the underlying causes. But more work must be done, and, even more important, farmers, industry and governments must act in the light of our present understanding of climatic change. Only then will we be able to face up to the changed world of the 1980s with confidence.

The best way to see how the problems confronting Britain, Europe and North America are related to the climatic problems of the rest of the world is to look first at how susceptible our industrialized society is to what are really quite small changes in climate. Then, moving to the overall situation, we can see what sort of changes are now in progress, and what the economic planners should allow for over the next ten, twenty or fifty years.

Agriculture, the most notorious sufferer in 1974, deserves first place in any investigation of the problems produced by our changing climate. In Britain, the situation in mid-June of 1974 was being described as 'the worst drought of the century'. Crops of grass for hay and silage were as low as 60 per cent of the 'normal' requirement, and dairy farmers were expressing concern about the effect of this shortage of food on their cattle, which must inevitably mean higher prices for milk and dairy products.

Soft fruits suffered severely, as did root crops, including potatoes. And one of the worst effects of the drought was on the sugar beet industry, with more than 15 per cent of the crop being ploughed in just at a time when world shortages were beginning to force sugar prices upwards and produce a shortage in the shops.

The new climatic pattern becoming established over Britain seems to be one in which the plentiful gentle rains traditionally associated with the island are being replaced by long, dry spells broken by violent storms. That provides some comfort at least for the authorities responsible for providing our water for household and industrial use. From the point of view of topping up a reservoir, a good strong storm can be just as good as a longer period of standard British drizzle. And, of course, long dry spells of settled anti-cyclonic weather could do wonders for Britain's tourist trade. But for the poor farmers – and that means for anyone who wants to eat – the story is much more gloomy.

Of course, Britain is fortunate in one respect; a drought in Britain is not the same as a drought in, say, Ethiopia. It would be quite possible for British farmers to change their habits and turn to other forms of agriculture, breaking with tradition but preserving their livelihood and keeping us fed. Provided the climatologists can give accurate guidance to the farmers about what to plant over the coming decades, there should be enough to eat. But the problem looks more serious the further afield we look.

In the U.S., the summer of 1974 was even more disastrous to agriculture. Described there as 'the worst drought to hit America since the 1930s', by mid-August it was estimated that farmers and ranchers had lost more than nine billion dollars in crops and livestock. The Agriculture Secretary of Iowa said even then 'it's wishful thinking ... to believe the drought isn't going to have an effect on overall production'; among crops worst hit were corn and soya beans, largely used as animal food.

Britain imports about a million tons of such food each year, much of it from America, and shortages increase the price of beef both in the U.S. and in the rest of the world. To cap this seemingly desperate situation, in the very week that the Governor of Oklahoma was calling for thirty-five counties in the south and west of his state to be declared drought disaster areas, the news from Venezuela was that drought in the state of Guarico had killed 20,000 cattle, stopped the irrigation of 144,000 acres of rice, and threatened the survival of 100,000 families.

So it is not even a question of Britain or America buying in traditional foods from outside. The global deterioration of climate means that we must get used to different foods entirely. One obvious possibility is to cut out the inefficient 'middleman' in protein production – the animals. There is no good reason why we should not get our protein direct from plant sources, such as soya beans, although it might be necessary to process this into artificial 'meat' to conform with eating habits ingrained people in the industrialized nations. But even plant foods are likely to be in rather short supply.

In the *Bulletin* of the World Meteorological Organization, an assessment of the impact of weather on grain supply drew attention at that time to the fact that huge reserves of grain which had existed in storage as recently as 1972 had been used up, and 'spare' land in North America has now almost all been returned to active production and there is virtually no spare agricultural land anywhere else. With no reserves of grain or land, the world is in the hazardous position of leading a hand-to-mouth existence from one crop to the next – a situation in which a spell of bad weather, let alone a continuing climatic trend, could have disastrous consequences. The sober pages of *The Economist* spelled out just what that might mean in an article on August 3rd, 1974: world grain stocks at their lowest for 26 years and a combination of floods in Canada and drought in the U.S. forced the price of wheat up by one-third fio $4.41 a bushel; the price of corn up by one-third to an 'unprecedented' $3.74; and soya beans up by nearly two-thirds to $8.64 a bushel. And these soaring costs contribute as much as 60 per cent to the costs of livestock farmers. With North America usually providing two-thirds of the world's wheat imports and animal feed, and virtually all of the world's soya beans, these figures really matter, whether you live in Birmingham, Alabama, or Birmingham, England.

Even so, for industrial nations such as Britain and the U.S. these worrying figures may soon pale into insignificance when the calculations of the effect of climatic change on industry and the home are worked out in the same detail – no one has done so yet, but the probable trend is easy to see. Weather and climate affect all aspects of our modern society. The effects of transport – for example, the choice of airline routes to take advantage of prevailing winds – are fairly obvious. Railway systems are likely to be less affected by severe weather than either airlines or road transport – but only if the railway system is prepared for such weather. An amount of cold and snow that causes little inconvenience to the railways of, say, Scandinavia or Canada, could wreak havoc in Britain. The point is, of course, that ice- and snow-clearing equipment is expensive, and the investment is only worthwhile if it is going to be used fairly often during the winter seasons over ten years or so. Effects of severe climatic conditions on industry are not only felt through the disruption of transport: the most basic effects can cause trouble – is a factory or office likely to be simply too cold to work efficiently in winter? (Or too hot in summer?) If there is a drought, where will factories obtain the copious supplies of water that are so essential to many industrial processes? And there are the effects on the work force. A bad winter can cause loss of production simply because more people are ill; a good summer can also be a problem if more people decide to take unofficial days off to take advantage of the sunshine. It's easy to imagine these common experiences multiplied by conditions outside the run of climate that we now accept as normal.

Here is an example of the importance of predicting climatic change. Two very severe winters have occurred in Britain over the past thirty years – one in the mid-1940s and the other in 1962-3. That frequency of severe weather is not enough to justify lavish expenditure on 'defensive' equipment for railways, or even to encourage expenditure on insulation for housing and lagging for water pipes. But if there is a strong likelihood of two or three such winters occurring over the next ten years then the situation is very different. Housing is an area where both the likelihood of frequent very severe winters and the prospect of a 'permanent', or at least long-term, small change in the average temperature should play an important part in planning. It is also an area where very little notice has been taken of climatic factors. Double glazing, insulation of cavity walls and roofs, and lagging of pipes seem obvious precautions against severe conditions. Such measures are now becoming more widespread in Britain and America because of the increasing cost of fuel. But we are still a long way from anything that could be called a common-sense attitude to preparing our houses to cope even with the weather we are used to, let alone with any deterioration.

Consider the possible impact of climatic change on the development of Britain's North Sea oil reserves. Suppose, for example, that the average temperature in Britain declined by about half a degree over the next ten years. This may not happen, but it is the sort of trend which can happen over that sort of time-scale, and there is, as we shall see in later chapters, certainly a hint of such a trend already in progress. At the same time, suppose that the weather pattern of the past three years or so becomes firmly established as the norm, with storms over the North Sea becoming rather more common than was predicted when exploration for oil began in that part of the world.

Clearly, as it gets colder more fuel will be needed for central heating systems and there will be a greater demand for electricity as we all try to keep homes, offices and factories at temperatures we consider comfortable, while heat continues to leak from our ill-insulated buildings. At the same time, more severe conditions in the North Sea are already hampering the job of getting the oil ashore, and even one really severe winter could have a dramatic effect on the date when Britain eventually starts to reap the full benefits of North Sea oil. The 'great white hope' of Alaskan oil is, of course, equally at the mercy of the weather.

So, by the mid-1980s, we could well find that instead of North Sea oil supplying most of Britain's needs, less oil than expected is flowing and at the same time demand has increased well beyond present forecasts. There is no need for detailed calculations to show how disastrous that situation would be for the balance of payments – the economic barometer which inhabitants of all the developed countries have learned to watch carefully.

That may not happen, but the important point is that such a scenario is plausible, and that account must be taken of such possibilities in long-term economic planning. The cost of the amount of research into climate which could lead to reliable prediction of the average weather for the next decade or so is probably less than that of one oil rig. We have already got past the situation where the present global climatic situation can be dismissed as just a lean spell, and there is no doubt that such an investment in research would be worthwhile. But to see just how things are going, and what might be done to counteract the worst effects of climatic change, it is not enough to take a parochial view. We must look at the workings of the whole atmosphere – 'The Weather Machine', to borrow the title of a BBC TV special. Only when we can see the whole picture clearly will be able to understand what is happening to our part of the worldwide jigsaw puzzle.

2 The Global Picture

To understand how climatic change is going to affect us, or any particular region of the globe, we need to look at the changing patterns of two main indicators – temperature and rainfall. And we must be very careful not to read too much into evidence gathered over only a few years, or in one restricted area. In Britain, for example, we have had a run of four fairly mild winters between 1970 and 1975; but that does not mean that the temperature of the world is increasing. In fact, those four warm winters are just a local and possibly temporary reversal of a worldwide cooling down.

The evidence of this decline in global temperature comes out clearly when averages are taken over five-year periods, using measurements from all over the world. In the U.S., the Environmental Data Service and the American National Center for Atmospheric Research have between them produced this kind of average covering the 100 years from 1870 to 1969. Until the early 1880s, the measurements 'only' cover the northern hemisphere, and show a fall in the average temperature. For the following ten years or so, the worldwide temperature (averaged over successive five-year periods, 1880-84, 1881-85, and so on) stayed more or less constant. But then things started to change quite dramatically.

From the 1890s right through to the mid-1940s, with only a slight dip around 1905, the global air temperature increased. Altogether, this increase amounted to about $\frac{1}{2}$°C. That doesn't sound a lot, but it meant that people gradually got used to slightly easier conditions; there were fewer severe winters and farming became easier. Of course, that fifty-year period was a time of great change in many other ways, so perhaps the temperature trend went largely unnoticed amidst the revolutionary changes sweeping the industrialized world.

But meteorologists noticed the trend, and by 1950 they were beginning to

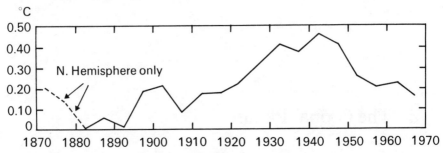

Fig. 1a Variation of global mean surface temperatures over the past century. These 5-year averages were obtained by Murray Mitchell and N.C.A.R.

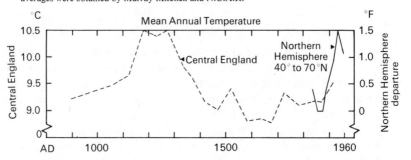

Fig. 1b Variation of mean temperature over the past millennium. Broken line represents 20-year means calculated by H. H. Lamb from historical data for England; solid line represents 20-year means for the northern hemisphere, calculated by Murray Mitchell

be worried by it. It was all very well to have 'better' weather – but what would happen if the world continued to heat up at such a rate? Ironically, an international conference was called to discuss this problem just at the time when, as we know, the temperature trend had reversed. Between the mid-1940s and 1970, global mean temperatures fell by about $\frac{1}{3}$°C; for the five-year period 1968-72, the average temperature recorded by the nine ocean weather ships which are stationed between 35°N and 66°N was more than $\frac{1}{2}$°C below the peak of the 1940s, and this local cooling continued in 1973. In worldwide terms, we are in a situation where the Earth is cooling more quickly than it warmed up earlier in this century.

What does this mean in local terms? Taking an average of the figures for England alone, the decade of the 1960s was about $\frac{1}{3}$°C cooler than the average for the preceding thirty years. The run of mild winters has recently put the overall average (for 1971-3) back to the previous level – but this average masks the fact that, while winters have been warm, springs and summers have continued to get colder. It looks very much as if we have been lucky with those winters, and we may well be due for a winter which will be much colder. If we get such a winter, with temperatures in line with the new average summer temperatures, it will come as even more of a shock than if we had been able to get used to the gradual decline in temperature over the past three or four years. According to Professor Hubert Lamb, the head of the Climatic Research Unit at the University of East Anglia, the size of the decline in temperature over England could mean a difference in the demand for fuel for heating of some 20 per cent over a 10-year period. Looking at the temperature figures for individual years for the past three centuries, he says that the heating requirement in the coldest year would be just double that in the warmest year since 1670.

Agriculture is very sensitive to changes of even $\frac{1}{2}$°C in average temperature. Harvests in England during 1960-73 were completed, on average, about 9 days later than during the preceding 20 years; this is almost entirely because the length of the growing season increases as the temperature falls – it simply takes longer for crops to reach maturity.

So the global decline in temperature is affecting Britain, even though we have had a run of mild winters. One great danger, of course, is that meteorologists and others might take alarm over this trend and take action just at a time when it might be reversing, echoing in reverse the situation in 1950. But there are two reasons to think that this time the trend might be more permanent. First, the 50 years from 1900-50 were very unusually warm. Professor Lamb has said that in climatic terms this was the most abnormal 50-year period during the last 1,000 years. And secondly, the present changes in climate can be linked to measurable physical effects – changes in the whole pattern of the global atmospheric circulation – which show no sign of reversing again in the near future. These circulation changes became clear to climatologists only in the past couple of years, when continuing droughts in the Sahel region of Africa hinted that there had been a real shift in the rainfall belts of the world.

There is so much evidence from various quarters to show that the climate is changing that meteorologists have difficulty in formulating a world picture. The simplest way to present the overall picture is to look first at one part of the world – the monsoon belt stretching through Africa and into northwest India for example – and to see how one meteorologist in particular related the climatic changes there to the global picture.

Dr Derek Winstanley, a London-based meteorologist, was studying changes in the rainfall pattern over North Africa as long ago as 1969. Then, he was working with the Anti-Locust Research Centre, where there was great concern about freak floods which had provided ideal breeding grounds for the locust. It has now become clear that these floods were caused by rains which were not freak at all, but were part of a new weather pattern which is becoming established. So just what did happen in September 1969 to cause the floods?

The immediate cause of the rains was that a large depression, or cyclonic disturbance, moved into the area, bringing its associated low-pressure troughs. Normally, these troughs and their weather systems move with the main belt of westerlies to the north of the North African coastline. The rainfall in September 1969 was unusual − it 'belonged' further to the north. but in view of other events we now know that it was not just one storm which had drifted further south − the belt of westerly winds itself has begun to move southward.

One way of looking at the changing situation with regard to the overall atmospheric circulation is by comparing the two extreme cases which can occur under 'normal' conditions. Derek Winstanley has done this, and he has worked out just how much variation in rainfall could be regarded as normal.

The key to this study is a comparison of so-called 'strong' and weak' circulation patterns for the northern hemisphere. According to Dr Winstanley, the strong type of circulation (see Fig. 2) corresponds to a situation when the oscillation of the westerly winds that sweep around the North Pole is confined within a narrow range of latitudes. In meteorological jargon, 'the circumpolar vortex is contracted.' What this means in practical terms is that the depressions (troughs) and ridges (anticyclones) which travel from west to east across the Atlantic produce changeable weather and high rainfall over the British Isles. The troughs in the circumpolar westerly winds do not push very far to the south, so that by and large the rainfall over the Mediterranean and the Middle East is relatively low.

Looking further south, with this pattern of circulation the tropical wind patterns extend well to the north, and allow the monsoon rains to reach high latitudes. This kind of strong circulation seems to be associated with a general warming of the northern hemisphere.

In terms of the problems of the present-day climatic situation, the strong circulation pattern for the northern hemisphere corresponds to high rainfall in Britain, in the Sahel region of Africa and in the monsoon region of India. At the same time, rainfall in the Mediterranean and the Middle East is fairly

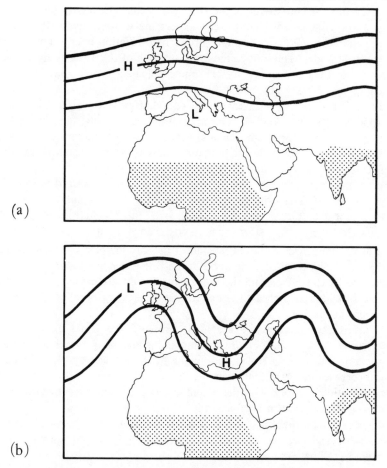

Fig. 2 Diagram indicating principal differences between extreme circulation types and associated rainfall patterns. (a) Strong zonal circulation produces high rainfall over Britain and allows monsoon rains to extend well to the north in India and Africa; (b) weak zonal circulation leads to low rainfall over Britain and suppresses the northward extent of the monsoon rains. (H = High rainfall; L = low rainfall; the change in the boundary of the dotted area shows the change in the extent of the monsoon region) (Dr D. Winstanley, *Nature*)

low, and there is in general a northward shift of the principal climatic zones. According to Dr Winstanley and to Professor Hubert Lamb, this strong circulation pattern was typical of the centuries just before A.D. 1200 and of the period from about 1700 to 1930. But now we are moving rapidly towards the opposite extreme, the so-called 'weak' circulation pattern (see Fig. 2(b)). That means that we can expect the westerly winds that circle about the North Pole to be more prone to oscillation over a fairly wide range of latitudes. This brings the main strength of the westerly flow of air further to the south, or, as the meteorologists would have it, 'the circumpolar vortex is expanded.'

In such a situation, the weather systems that move towards Britain and Europe across the Atlantic Ocean move more slowly, and persistent anti-cyclones (high-pressure systems) have a chance to become established. When that happens, the depressions (low-pressure systems) moving eastward are 'blocked', and are deflected to the north or to the south. For Britain, this means that the rainfall and the occurrence of westerly winds are both decreased. The troughs can now extend down into sub-tropical and even tropical latitudes, and the rainfall over the Mediterranean and the Middle East increases, just as has been observed in recent years.

But the influence of the 'expansion of the circumpolar vortex' does not stop there. The tropical circulation in the belt just north of the equator is squeezed by this effect, and the monsoon rains cannot extend so far to the north. In addition, such a situation corresponds to a general cooling of the northern hemisphere.

That is the situation we are in at present. The zonal circulation is decreasing in strength, rainfall over England is becoming less, and there are associated increases in the rainfall of the Mediterranean and Middle East, and decreases in the rainfall over the Sahel, Ethiopia and northwest India. The climatic zones of the northern hemisphere are shifting southward, much as was the case between 1300 and 1700.

This is a very simplified explanation of the problem of climatic change. But, as we will see, other pieces of evidence fit very convincingly into this simple overall pattern.

By 1973 Dr Winstanley was able to piece the picture together more clearly. As long ago as the 1930s, there were some signs that the climatic zones of the world were being squeezed towards the equator, which for us in the northern hemisphere means that rainfall belts are moving south. Although this now looks like a continuing trend, it has not been a cause for concern until now because some factor (as yet not identified for certain) caused a temporary halt in this drift in the 1950s. But through the 1960s, it seems that the trend has continued even more rapidly, almost as if it is trying to make up for lost time.

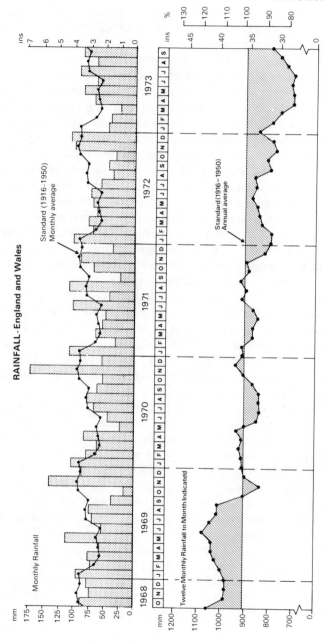

Fig. 3 Top diagram shows average monthly rainfall for England and Wales. Bottom diagram shows rainfall in England and Wales, 1968-73. The horizontal solid line denotes the average for 1916-50. According to some meteorologists, this decline in rainfall over Britain is related to the changes in the sub-tropics, and over the Mediterranean and the Middle East, during the same period. (*Annual Report of the Water Resources Board*, 1974)

But I am moving on too fast. To understand these changes it is necessary to have some idea of what produces the different climatic zones – the belts around the Earth, which are centred on different latitudes, and which have much the same climate. Two basic processes drive the weather systems: one is that the equator is hotter than the poles, so warm air tends to rise at the equator and fall at the poles before returning as winds over the Earth's surface to complete the pattern of convection. The other is that the Earth is rotating, so there is a tendency for winds to blow "sideways". These two affects combine to produce the spiral winds which characterize low-pressure regions (cyclonic disturbances) and high-pressure regions (anti-cyclones). In low-pressure systems, the winds spiral inwards to fill up the low, but anti-cyclones are like domes of high pressure, with winds spiralling outwards 'downhill'. If the Earth were a smooth sphere covered with water, that would make weather forecasting very easy. But mountains and land masses deflect the winds, and the different heating of land, sea, forests and snow combine to complicate the task of the meteorologists.

All we need bother about is the fact that these effects do produce climatic belts, and the prevailing winds which have proved so useful to the old clipper ships and to modern jetliners alike. One of the keys to this system of climatic belts is that a large area of high pressure is centred on each pole and surrounded by a circumpolar region of low pressure (Fig. 4). If this 'polar high' expands or contracts, the whole of the climatic belt system of a hemisphere is squeezed or stretched accordingly – and as far as we know at present, when these changes occur in the northern hemisphere they are mirrored in the southern.

Returning from generalities to the specific case of the monsoon in recent years, we can see how Winstanley's study points the way for an understanding of climatic change over the whole of the northern hemisphere. Typically, rainfall occurs when warm, moist air from the oceans rises over land, especially over mountains. The monsoon – in India, say – occurs because from March to June each year the land is heated by the growing warmth of the Sun, so that the hot air rises and produces a great area of low pressure into which the moist air from the sea to the southwest must flow. At least, that is how the pattern used to be. But now, after half a decade of failures of the monsoon, it is clear that a new pattern is established: the rainclouds seem to form, but they are blown away south of what have now become drought areas. For a chain of meteorological stations stretching across Mauritania, Mali, Niger, Sudan and northwest India, the rainfall averages declined by about half between 1957 and 1970.

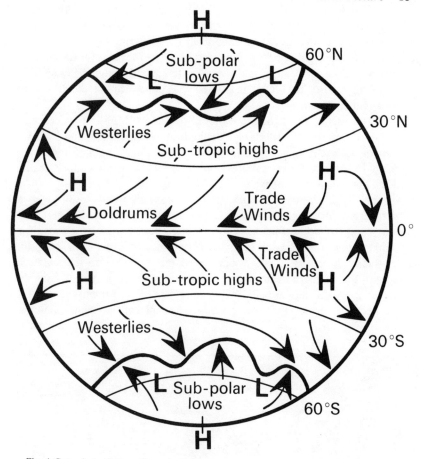

Fig. 4 General circulation and sea-level pressure distribution of the atmosphere around the world.

Strictly speaking, these are rainfall averages calculated as running means for the 5-year periods centred on those 2 years, like the 5-year temperature averages already mentioned. So they provide a much better guide to long-term trends than figures for any one year in isolation. The pattern of climatic change in the zone just below the Sahara (the Sahel region) is similarly one of declining rainfall, which Dr Winstanley links to other changes further north.

The schoolboy mnemonic for remembering the characteristics of a so-called 'Mediterranean' climate is 'warm, wet westerlies in winter'. But it now looks as if in the Mediterranean itself those warm, wet westerlies are moving not across the northern Mediterranean, as they did for decades up to the 1960s, but across the southern Mediterranean bringing floods (and encouraging locusts to breed) in some areas, and helping the desert to bloom again in others.

In North Africa, north of the Sahara, and in the Middle East, rainfall measured during the late 1960s was the highest ever recorded, and it shows no sign of declining again. Ironically, just when the droughts are causing hardship in the south, agriculture is being given a better chance in the northern margins of the Sahara.

All this evidence comes from studying rainfall trends only over a restricted part of the globe, and only covering a couple of decades. The obvious thing to do, before anyone can use these trends as a basis for action, is to look further back in time and over a greater area of the world. When this is done, the situation looks even more dramatic – the climatic changes of the northern hemisphere may, it seems, follow a cycle of roughly 200 years in which rainfall belts shift regularly to the south and then north. If this theory is correct, then the southward advance of the Sahara, and the

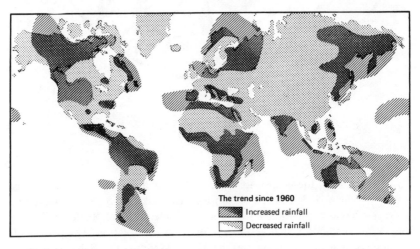

Fig. 5 Approximate trends in rainfall around the world during the 1960s. Rain which falls over the sea is essentially wasted, at least as far as farmers are concerned. So although the same amount of rain fell around the world in 1970 as in 1960, it did less good in agricultural terms.

related changes in climate which are affecting Britain, the U.S. and the whole world, will continue for another 60 years.

Detailed rainfall records for the regions of Africa and India studied only go back to about the beginning of this century. But these records testify that the climatic changes in different zones are linked to one another. In the Middle East and Africa north of the Sahara the wet season extends from October to May. From the beginning of this century, that winter-spring rainfall first decreased, on average, to reach a minimum in 1926-7. Since then, the rainfall has increased. Simultaneously, the monsoon rainfall of the Sahel south of the Sahara, and in the northwest region of India, increased to a maximum between 1925 and 1927. Since then the monsoon rainfall has declined, except for the temporary 'hiccup' of the 1950s.

We don't have much information about how rainfall varied in these areas before about 1900, but in Britain meteorological records have been kept for hundreds of years. Rainfall in Britain reached a peak around 1930 and has since declined in more or less the same way as it has south of the Sahara. It does look as if all these changes are part of a global pattern; historical records from Britain going back over a thousand years can be used as a guide to climatic changes also.

Of course, rainfall records are not accurate for the whole millennium. But there are historical records of droughts, great floods, hard winters, and so on, which is enough for Professor Lamb to have produced a rough guide to the trends over that time. It seems that rainfall in various parts of Britain peaked in the middle of the millennium, declined until about 1700, and then rose again to a peak in 1930. That is barely sufficient evidence of a 200-year cycle of climatic variation. Even if the cycle is not strictly periodic, however, it does show that trends over that sort of time-scale do take place. The immediate problem is to find out whether the present trend, established in the late 1940s, is such a long-term trend, or just a minor 'wobble' in the worldwide climate. Statistical tests indicate a 200-year climatic cycle, and very recently, as we shall see, a physical theory to account for such a variation has been developed.

If the trend is real, then the implications for the Sahel are clear. By the year 2030, when the 200-year trend might be expected to reverse again, the Sahara desert will extend some 100 kilometres to the south. Of course, there will have been a corresponding southward retreat of the northern fringe of the desert, but that will be small comfort to the displaced nomads of the Sahel region.

And what would all this mean to, say, Britain and other countries further north? This was well described in what now seems a prophetic article in the journal *Weather* in August 1970, by Dr A. H. Perry, of University College, Dublin: 'The decade of the 1960s may well be remembered as the first when climatologists became convinced that the climate over large parts of the northern hemisphere was experiencing a reaction to the genial conditions of the earlier part of the century.' He went on to say that 'the very severe winter of 1963 over large parts of Europe, the excessive amounts of ice in the polar seas in 1968 and the decline in frequency of westerly winds ... have been the most spectacular manifestations of more inclement conditions.' And it now looks as if those conditions will get increasingly inclement for 50 or 60 years. The need, clearly, is to find out just when the trend will be reversed and how severe conditions are likely to become. To answer such questions, we need both historical and geological records of previous trends of this kind, and the latest observations and physical explanations of what is happening to the world's weather today.

3 Previous Climatic Changes

In order to find patterns of climatic change that apply over tens of years, and which we might use to predict how the climate will change between now and the year 2000, we need to look at evidence covering hundreds or thousands of years. Five years of cooling, like the period 1968-73, mean very little in themselves. But if we can find that periods of cooling in the northern hemisphere occur regularly every few hundred years, and that the five-year trend in 1968-73 fits into such a longer term pattern, then we might discover how long we can expect the present cooling to last.

The situation is much the same on a smaller scale. If it rains on Monday, that doesn't mean that rain always comes on a Monday. But if, for example, a survey showed heavy rainfall on the second Monday of every month over a year or so then it might be worth remembering to carry an umbrella when the second Monday of a month came around again!

The snag, of course, is that the patterns we want to find cover extremely long periods. We need data covering thousands of years at least, preferably tens of thousands of years. There is, in fact, a good deal of historical evidence, particularly for Britain and Europe, and this can be used to fill in details of the climatic picture, but these details are of little use without evidence from other parts of the world. And for global information covering thousands of years there just isn't any historical evidence available; however, geologists can tell a great deal about past climates from evidence preserved in sediments. And 'palaeobotanists' can now infer remarkable detail about climatic change in many parts of the world from studies of the ring patterns of ancient trees — the relatively new science of 'dendro-climatology'.

Tree rings provide an invaluable source of palaeoclimatic information because the amount of growth of a tree in any one season depends on the

19

temperature and rainfall during that season. In principle, the technique is very simple. A sample from an old tree can be analysed simply by counting the rings from the outside inwards, to give the year corresponding to each growth ring, and then plotting the thickness of each ring on a graph to show how climate has varied. But in practice the analysis is more difficult.

The main reason for this is that the rings are affected by *both* rainfall *and* temperature. So does a thin ring imply a very cold growing season or a very dry one? It is far from easy to resolve this problem, and in particular the frequency of variations of both rainfall and temperature in Europe makes it difficult to use European trees for accurate dendro-climatological studies. But in some parts of the world the climatic pattern is less confusing, so more information can be gathered from tree-ring studies.

To some extent, this kind of work is more of an art than a science. The best dendro-climatologists (and there are very few of them) develop an almost instinctive 'feel' for the evidence contained in the tree-ring record. With the aid of this instinct and a great deal of very patient work studies of trees on the North American continent − in particular, the long-lived bristle-cone pine − have been used to build up a picture of climatic changes in the area over very long periods.

One particular study, carried out at the University of Arizona, shows the principal features of climatic change over the past 850 years in the plains and plateaus of the U.S. and Canada. Throughout the thirteenth century, it seems, North America was drier than it is today; in the fourteenth century this drought gave way to a long stormy period, but by the end of the sixteenth century, dry conditions were back − indeed, the period 1575 to 1600 was the driest 25 years of the past 7 centuries. About 1670, these fairly long-term changes were replaced by a completely new pattern of climatic fluctuations, with changes occurring every 20 or 25 years, and this pattern has continued to the present day, with drought years occurring just before 1900 and 1930 as part of the pattern. It's not unlikely that the present droughts in the U.S. are also related to these short-term fluctuations.

Tree-ring evidence helps climatologists to understand variations in the general circulation of the atmosphere over thousands of years; but it is not used in isolation, rather, it is used along with other evidence to build up a reliable picture of climatic changes from many small pieces of evidence.

One of these other pieces in the climatic jigsaw comes from the studies of pollen deposited in sediments. This provides a clue to the kind of plants common when the sediment was laid down, and thus indicates the climate; another clue comes from studies of variations in the chemical composition

Fig. 6 Anomalous pressure variation during winter that persisted from A.D. 1816-45 as reconstructed (predicted) from anomalous variations in the widths of tree rings (units are calculated as millibar departures from the mean pressure for 1899-1939, 1945-62 and averaged by pentads). Dots mark those departures that are twice the residual standard error. The period from 1816-30 is characterized by a weakening of the Aleutian low (higher than normal pressure) and a weakening of the high over the sub-tropical Pacific, 25° N latitude. There is a strengthening of lows over Newfoundland and Hudson's Bay in 1826-30 and again in 1836-40. The anomalies are less marked in 1816-20, 1831-5, and 1841-5. (Professor Hubert Lamb).

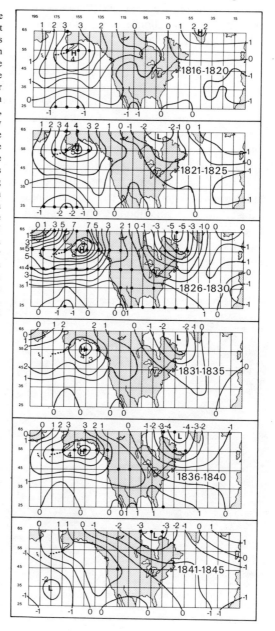

of ice laid down in the Artic and Antarctic. But one of the best indicators of periodic changes in climate comes from a source which is remarkably akin to the study of tree rings.

Just as trees 'deposit' a layer of new wood each year, so certain lakes and ponds deposit clear-cut layers of silt and clay each year. These layers, called 'varves', can be traced in samples of sediment drilled from the bottom of a lake which freezes each year. They are produced because during the cold season no new material enters the lake, which is frozen over, and only very fine sediment from the waters of the lake is deposited. In the spring, the ice melts and fresh water brings in a flood of new sediment. The result is a varve layer which looks just like the layering of tree rings. (Fig. 7).

No two successive years produce the same thickness of sediment, and the varves provide a good guide to climatic change – in particular, the strength of the spring thaw and flooding is very important in determining the amount of silt deposited. In Scandinavia, varve records covering almost 14,000 years have now been pieced together. Like tree rings, they can be dated by counting the layers; and varves can even be related to the tree-ring evidence,

Fig. 7 Sediment sample obtained from an old lake bed, showing characteristic layering of sediments. These varves provide an insight into past climates since the thickness of each layer depends on the weather when it was laid down.

by comparing patterns of thin and thick varves with patterns of thin and thick tree rings.

Once again, there is a problem of interpreting the varve record in climatic terms: does a thick layer of sediment imply high temperatures and rapid thawing, or high rainfall and flooding?

When the variations in thickness of varves and tree rings over thousands of years are analysed statistically, several patterns emerge. One of these suggests a link between sunspots and climate, since the records show variations with a period of around 11 years, similar to the period of the 'sunspot cycle'. Another strong, recurring period seems to be close to 90 years, and there is a suggestion of another at just under 200 years. These cycles are important clues to the nature of climatic change in themselves, as we shall see. But just how important they are, and have been over not just thousands but billions of years, emerges from studies of 'fossil' varves.

Once a varve layer has been deposited at the bottom of a lake, it is subject to the same geological process as any other sediment. In other words, over geological periods of time the varves are compressed into rock. The rock strata containing the varve traces may then be moved by the geophysical processes which mould the Earth's surface, and might end up as part of a mountain range, or be exposed at the surface. In one case, traces of a 'fossil' varve record covering 900 consecutive years were found in shales from Alberta, Canada. These shales are dated as Upper Devonian – about 300 million years old. So that particular varve record reveals climatic variations over a period of nine centuries 300 million years ago.

This is astonishing enough in itself. Of course, the varves only record some average of the effect of temperature, rainfall, storminess, and so on. But statistical tests can show how the weather varied, on average, from year to year during the best part of a millennium in the Upper Devonian. And the result of the statistical tests is that exactly the same periods – 11, 90 and about 200 years – were important in climatic changes even then. Clearly, such changes must be linked to the basis of atmospheric circulation and the fundamental nature of climatic change on Earth.

All this is useful background to understanding the climatic problem, but what can we learn from past situations which are comparable with what is happening to the world's weather today? Once again, the geologists, this time in collaboration with archaeologists, can provide important clues.

The river Nile has long held a fascination for scientists from different disciplines. It is particularly useful as an indicator of rainfall variations, since it gathers its waters from an area large enough to be affected by climatic changes, rather than just by the more local weather fluctuations. It has been known since the end of the nineteenth century that the flow of the Nile has varied considerably over thousands of years. But this knowledge, rather than providing clear-cut climatic information, was for many years the subject of fierce controversy. The main point at issue was: did the periods of glaciation in Europe and East Africa – the ice ages – correspond with increased Nile flow? Or were the glaciations contemporary with a period of aridity in the tropical regions where the Nile rises?

These questions, which caused sometimes bitter arguments among climatologists, geologists and others in the early part of this century, now assume a more urgent topical importance. For it seems possible that the droughts in the Sahel and in Ethiopia might be just the kind of climatic change which could cause a decline in the Nile flow. If so, can we expect a period of increased glaciation – a mini ice age?

In 1974 Dr M. A. J. Williams and Dr A. Adamson, of Macquarie University in New South Wales, Australia, published the results of a new study of this problem. This topical study focused attention on the Nile as a climatic indicator, and provided some of the best evidence yet to relate Nile flow changes to changes in ice cover of the northern hemisphere.

Surprisingly, it was only as recently as 1962 that the first piece of direct evidence was produced to show that the glaciers of Ruwenzori in East Africa were on the wane about 15,000 years ago, at the same time that the glaciers of Europe and North America were retreating at the end of the latest phase of extensive ice cover. By the mid-1960s, geological evidence had also been gathered to show that other East African glaciers – such as those of Mount Kenya and the Kendall – were also retreating at the same time. So, although it might have been expected that African glaciers would follow the pattern of advance and retreat known for glaciers further north, it is only in the past few years that climatologists have been able to establish for certain that the glaciation of East Africa is directly related to the overall ice cover of the northern hemisphere.

During the discussion of these problems early on in the present century, some climatologists argued that if ice caps were formed over East Africa, the monsoon winds would be displaced to the south. In that situation, rainfall over Ethiopa would be reduced, and there would be a corresponding reduction in the flow of water entering the Nile proper from the Sobat, Blue Nile and Atbara tributaries, which today account for more than 90 per cent of the main Nile flood.

As long ago as 1914, the idea was put forward that when the northern

Sahara was wet the southern Sahara was arid and extended further to the south. In that case, the arid belt might be as much as 5° further south during an ice age, with a 'weak' atmospheric circulation and a lower overall temperature in the northern hemisphere – and presumably in the south as well.

That is a rather familiar picture. It bears a striking resemblance to the ideas put forward by Derek Winstanley to explain present climatic changes affecting all of the northern hemisphere, not just the Sahel and other parts of Africa. Some people have been led by this similarity to jump to the conclusion that present-day circulation and temperature trends are on the way to producing an ice-age situation.

Several theories have been put forward to explain Nile-flow variations; I find the evidence put forward by the Macquarie University team the most convincing. That is partly because it is the most up to date, and so should be more reliable. But these are early days yet for the study of global climatic change, and though the evidence is disturbing it is possible that discoveries yet to be made will lead to a lightening of the gloom.

One reason why studies of the Nile flow have been so confusing is that people have generally tried to explain the flow of both the White Nile and the Blue Nile within one general scheme. But the rivers are very different. In 1866 Sir Samuel Baker described the Blue Nile as a mountain stream rising and falling with great rapidity, and the White Nile as a river of lake origin flowing through a vast treeless swamp. These graphic descriptions are still valid; it is hardly surprising that attempts to explain fluctuations in both rivers by one theory have met with difficulties.

The study by Drs Williams and Adamson in 1973 concentrated on the White Nile alone. This concentration on just one river makes the puzzle a lot easier to resolve, and that is another reason why the 1973 study seems one of the best guides yet to how Nile flow relates to climatic change. Working with Dr J. D. Clark and an archaeological team from Berkeley, the Australians collected freshwater shells from sites several metres above the present-day flood level of the White Nile. These shells can be dated by the radiocarbon technique to indicate just how long ago it was that floodwater carried them to their present positions. The team also looked at soil samples from layers at different depths near the banks of the White Nile today. Again, the sediments show evidence of periods of flood or drought.

These two different lines of attack gave unambiguous results: in general, the climate was dry from before 14,500 years ago until about 12,000 years ago; at that time the White Nile flow was much less because there was little outflow from the Uganda lakes. Indeed, the region was so arid that the White Nile may have failed to reach the main Nile. About 12,000 years ago, there was a rise in the level of the river, corresponding to overflow from Lake Victoria, and the resulting flooding has left its mark in the form of alluvial clay in the sediments near the present river. This coincides with the end of the most recent ice age. The level of the river stayed high, according to the evidence presented by Drs Williams and Adamson, for 8,000 years, falling again about 4,000 years ago.

This evidence – that when the river is low, the ice cap is more extensive – provides strong support for the idea of a link between ice ages and periods of drought in Ethiopia; by implication, it ties in with the idea that ice ages are associated with weaker atmospheric circulation. But it provides no answer to the chicken and egg question of whether increased ice cover causes weaker circulation, or whether weak circulation, once established, leads to an ice age.

Other archaeological evidence also shows that the time of glaciation in Europe and North America can be related to climatic changes throughout the northern hemisphere. In India, for example, it seems that the Great Sand Desert in the northwest has expanded and contracted two or three times over the past few thousand years, and that these changes might be associated with climatic changes further north. The area where the desert seems to advance and retreat lies close to the present border between India and Pakistan. Today, the eastern margin of active natural sand dunes lies roughly between 25°N 71°E and 29°N 74°E, running more or less northeast – southwest. But in the past the desert may have reached 500 kilometres further south and east – as far as Ahmadabad on the coast and Jaipur and Delhi to the northeast – although, of course, that was long before those cities were built.

Once again, the approach used by archaeologists and geologists to study the variations depends on obtaining samples of soil and fragments of man-made objects – pots and the like – from different layers below the present ground surface in the region. The depth of a sample indicates its age, and the nature of the soil or sand found at any depth indicates the climate prevailing when that layer was deposited. In addition, the presence of man's artefacts is a pretty reliable indicator that enough water to live on must have been available at the time the artefacts were made.

One detailed study of this kind was carried out in the area by Dr A. S. Goudie, of the University of Oxford, and Dr Bridget Allchin and Dr K. T.

M. Hegde, of the University of Cambridge, and was reported in the *Geographical Journal* during 1973. Their evidence points to an interval of 'significant wet climatic conditions' separating two 'phases of aridity'. Assuming that rainfall of less than 250 mm. a year allows active dune formation, the Oxbridge team was able to provide a fairly accurate guide to past climates in the area (present-day rainfall in Ahmadabad is about 780 mm. a year, and in New Delhi it is about 660 mm. a year).

The evidence suggests a moist phase from about 8000 B.C. to 7500 B.C., a drier phase from 7500 B.C. to 3000 B.C., and a wet phase from 3000 B.C. to 1800 B.C. The Oxbridge team does not, in fact, believe that these changes in climate thousands of years ago necessarily have any relevance to what is happening in the same part of the world today; they believe that the present-day desert encroachment is entirely due to overgrazing and bad farming practice by man. Certainly, the Indian evidence alone is no cause for alarm. But when it is related to other evidence – the Nile-flow variations, for example – the problem looks more disturbing.

What we need to find is a factor that we know, from archaeological and geological evidence, indicated climatic change and that has also recently changed. News came in 1974 of the discovery of just such a change – the ice and snow cover of the northern hemisphere. This is perhaps the key part of the understanding of the importance of climatic change today; without this clue there is circumstantial evidence, but with it there is concrete proof that the climate is changing. On that basis alone, the discovery of these ice- and snow-cover changes certainly warrants a chapter of its own.

4 Ice Cover and Global Weather Shift

Changes in the overall size of the Arctic ice cap are of great importance to the climate of the northern hemisphere. But before the advent of satellites, which could monitor such large-scale effects, meteorologists had no data relating such changes to changes in weather and climate. Indeed, although the first man-made Earth satellite was launched in 1957, it was 10 years before satellites became sophisticated enough to report how ice cover changes with the seasons on a global scale. Today, there are several satellites in orbit which can send back to Earth pictures of the overall appearance of the northern hemisphere. Pictures of ice cover are produced by an ordinary photographic process, and also by infra-red scanning.

But the infra-red scanners are not needed to pick out the changes in ice cover which Drs George and Helena Kukla found by looking at a five-year series of weather maps kept in the files of the U.S. National Oceanic and Atmospheric Administration (N.O.A.A.). In February 1974 they published the remarkable results of their survey in the journal *Science*. With the caution of the professional scientist, their article carried a straightforward – indeed, almost dull – title: 'Increased Surface Albedo in the Northern Hemisphere'. But the sub-title of the article was more exciting: 'Did satellites warn of the weather troubles of 1972 and 1973?' And the authors answered their own question in the first paragraph of the article, which summed up the situation so well that it is worth repeating almost in full.

Snow and pack-ice cover in the northern hemisphere formed earlier in the year and covered a larger area in the past 3 years than it did 7 years ago, when systematic satellite mapping began ... The difference was most pronounced in the fall and was especially large in 1971. The anomalous global weather patterns of 1972 and 1973 may be the result of these developments.

That must surely be one of the most important paragraphs ever written during the course of the scientific study of climatic change. But the story behind that paragraph is one of careful scientific investigation and analysis; there is nothing half-baked or overdramatized about this work, and its implications must be taken most seriously.

The N.O.A.A. charts that were used in this study are basically weekly weather maps which are produced from photographs obtained by weather satellites. They show any snow or ice fields which last for 5 days or more, and they cover the entire region of the northern hemisphere where such cover occurs. And, as George and Helena Kukla point out, the location and duration of snow and pack-ice fields is the most important factor affecting the Earth's 'heat balance' over the seasons. This is because snow and ice are such effective reflectors of the Sun's heat. The radiation (heat and light) from the Sun which reaches the Earth is termed, in meteorologists' jargon, 'isolation'. Normally, vegetated ground reflects some 15 to 20 per cent of insolation, and calm seas reflect only 5 to 10 per cent. But snow-covered grasslands, or pack-ice on the sea, reflect about 80 per cent. The atmosphere is not heated directly by the Sun's rays passing through it, but by the warmth of the ground or sea underneath. So the reflected light and heat is completely lost into space – as the Drs Kukla put it, 'this represents a deficit in the Earth's energy balance.' And, in addition, pack-ice divides the atmosphere from the warmer ocean waters which could otherwise heat up the cool air.

There are other ways in which snow and ice affect the weather. The most important of these is that it takes a great deal of heat just to melt the frozen water when spring comes around. Just about 80 calories of heat are required to melt each gram of water – and it only takes 100 calories to heat 1 gram of water from freezing point to boiling point. Because of this, the greatest air temperatures in the mid-latitudes of the northern hemisphere occur about 6 weeks *after* the peak of the summer insolation, because so much of the Sun's heat goes into melting snow and ice.

So variations in snow and ice cover are indeed important for weather and climate. Just how then, has this cover changed since the end of 1967? In 1968, the minimum area of the globe which was covered by ice and snow duing the year was 38 million square kilometres. (That was measured in August 1968.) During the same year, the maximum cover reached by the snow and ice was 76 million square kilometres – just twice the minimum – and that was reached at the end of December 1968, and persisted for the first few days of 1969. To put these figures in perspective, the maximum cover (76 million square kilometres) was equivalent to 15 per cent of the surface of the globe.

But these global figures are deceptive, because there is a great difference between the two hemispheres. In the north, the polar regions are basically covered by a landlocked sea, the Arctic Ocean, but the south polar regions are covered by the southern continent of Antarctica. This produces important differences in the development of snow and ice cover in the two hemispheres.

The permanent ice cover in the north is about 10 million square kilometres, and in the south about 14 million square kilometres. But on the other hand, seasonal snow and ice cover reaches 50 million square kilometres in the northern hemisphere, and only 20 million square kilometres in the south. This is because the large area of land in the north provides a base for even thin snow cover to lie on. Most snowfall in the southern hemisphere falls on the sea and melts.

Over the 7 years from 1967 to 1973, the snow and pack-ice cover of the northern hemisphere showed just the kind of seasonal variations that you would expect: the minimum cover always occurred during the last week of August and the first week of September. The decline of the snow and ice fields generally began in February, and was very fast in April and May of each year, but slower in July and August, when most of the snow and ice had already melted.

Before the satellite photographs can be used to interpret changing weather patterns, however, these observations must be put on a solid basis of measurement. So the meteorologists have defined four snow-cover seasons (S.C.), which are similar to, but slightly different from, the four usual seasons. They are defined, quite straightforwardly, in terms of how much ice and snow there is at different times in the year. Looking at just the northern hemisphere: S.C.-summer is the period of the year when less than 15 million square kilometres of the ground and sea are covered by snow and ice; S.C.-winter, by contrast, is the period when more than 55 million square kilometres are covered. Logically enough, S.C.-autumn is the period when snow and ice cover is building up from 15 to 55 million square kilometres, and S.C.-spring is the period when the cover is declining from 55 to 15 million square kilometres.

During the 7 years of the recent survey, S.C.-spring was, on average, about 7 weeks longer than S.C.-autumn. S.C.-summer and S.C.-winter, however, were the same length. In 1968, for example, S.C.-spring started on March 9th, and lasted for 118 days; S.C.-summer lasted from July 5th to

October 10th (97 days) and S.C.-autumn from then until December 14th, just 65 days. Finally, S.C.-winter ran for 86 days, until March 10th, 1969, when the cycle began again.

But even this very simple analysis shows some interesting changes in the pattern from year to year. From 1967 to 1970, S.C.-autumn and S.C.-winter started earlier each year. And in 1970 something really dramatic happened.

The S.C.-winter which began on December 26th, 1969, lasted for just 35 days, until January 30th, 1970. That meant that S.C.-spring began that year before February had arrived – yet in all the other years S.C.-spring did not set in until March. In particular, in 1971, S.C.-spring began on March 21st, and that was the latest in any of the seven years studied. So there is a very striking difference between the pattern of S.C.-spring in 1970 and in 1971.

These very broad indications are reflected by more detailed studies which emphasize that something odd happened between 1970 and 1971. When the figures for the ice cover in individual calendar months are compared, it turns out that the values for February, March, April, September, October and November are very different in the two years 1970 and 1971. Broad changes also show up when the cover in particular months is looked at throughout the seven-year period. In general, there is a recent tendency for more snow and ice cover in S.C.-autumn: in October 1972 the cover was twice as great as in the same month in 1968.

In their own words, the American climatologists report that 'the period on our record could be divided into two sections, sharply differing in their patterns of seasonal distribution of snow and ice: before 1971 and after 1971.' The important question now is, how can these changes be related to the weather? In particular, do they explain the dramatic changes in the weather and the overall circulation that have disrupted the economies of many northern hemisphere agricultural communities in the past couple of years?

As we have already seen, the key to what's 'wrong' with the weather lies in the change in the overall circulation pattern of the atmosphere. There is no need to have detailed records of temperature and rainfall variations around the globe in order to pinpoint when the weather made its change for the worse. Taking the 7 years of the ice- and snow-cover survey, the main circulation patterns of the years 1968 to 1971 were pretty close to what has come to be known as 'normal'. But the weather of 1972 and 1973 was 'anomalous'. The change in circulation corresponds exactly with the change in ice cover.

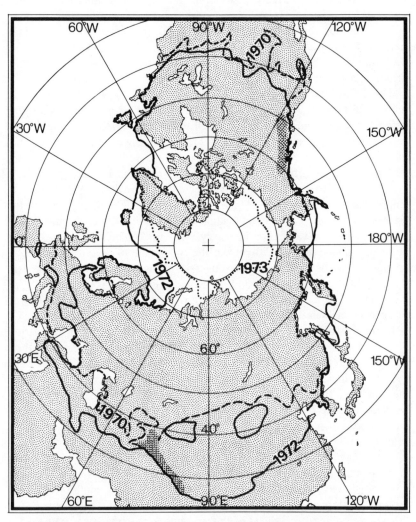

Fig. 8 Snow- and ice-cover boundaries for the northern hemisphere. The dashed line shows the boundary at the beginning of 1970 and the solid line the boundary at the beginning of 1972. The dotted line around the North Pole marks the maximum extent of pack ice in September 1973, covering large areas of formerly open water. Heavily stippled areas correspond to mountainous regions where snow failed to melt in the summers of 1971, 1972 and 1973. (Dr George Kukla)

Although this survey was only carried out for the northern hemisphere, and we have largely ignored the southern hemisphere in this book, in fact the change in circulation and weather was worldwide. During 1973, J. M. Bull reported some of the worldwide effects of the change, as it affected 1972, in the *Bulletin of the World Meteorological Organization*:

The general circulation of the atmosphere in 1972, over both the northern and southern hemisphere, differed considerably from the fairly consistent pattern that had prevailed each year from 1968 to 1971. In the northern hemisphere, the cyclonic activity over the North Atlantic and western Europe was much greater than previously owing to the fact that, after being quite weak for several years, the Icelandic semi-permanent low-pressure system intensified at about the end of 1971 . . . in the tropics and the subtropics, pressure was generally below normal and the tropical activity was in many areas very intense.*

To sum up some of the effects mentioned by Bull in more everyday terms: in 1972 the average temperature, especially in winter, was considerably below normal over all of the North American continent and over much of the North Atlantic Ocean and the Mediterranean. In the Baffin Island area, the average temperature over the whole year was a remarkable 4°C below normal. By contrast, much of the European Arctic Ocean experienced above normal temperatures. There were long dry periods both there and in western and central Europe. These droughts, as we have seen, affected agriculture seriously. As Winstanley has already noted, the Mediterranean and Middle East had above normal rainfall, as did the Pacific coast of North America and most of East Asia. And in the southern hemisphere, Australia experienced severe droughts during 1972.

Although the temperature over the North Pole itself actually increased slightly, over the rest of the globe the 'anomalous' weather of 1972 was accompanied by exceptionally large snowfields and by a record low in the air temperature north of 15°N. This was additional to the steady decline in the average surface air temperature of the Earth that had been going on for about 30 years. There seems little doubt that all these effects are related – but which is cause and which is effect?

George and Helena Kukla note that the shift in snow cover developed in

*W.M.O., Geneva, 1973.

1971, a year before the weather became anomalous; S.C.-autumn started three weeks earlier in 1971 than in any of the four preceding years. They say that 'it is still premature to say that the global weather pattern of 1972 was a response to the anomalous snow cover of 1971 ... but, in view of the close links connecting the distribution of snow and ice with heat reserves in the oceans and with atmospheric circulation, such a relation is probable.' They also mention that during the glaciation of a full ice age, the cover in the northern hemisphere must be between 60 and 70 million square kilometres. That is about 30 million square kilometres more than the present average; and in 1971 the increase in the average cover was about 4 million square kilometres. In other words, 7 such increases would establish a new ice age. Fortunately, it looks as if we are are not going to get 7 increases, at least not in successive years, since the average cover for 1972 and 1973 settled down around the new figure established in 1971. But the balance between present conditions and even a full ice age seems quite delicate.

Whatever the relationship between snow cover and atmospheric circulation, there seems little doubt that the real cause of changes in the climate must come from outside the Earth. A dramatic volcanic eruption – like Krakatoa – might put enough dust into the air to decrease the amount of heat reflected and thus cause a climatic change (see Chapter 8). So far we have been looking at how the different bits of the weather machine react and interact. To get longer-term warnings of how the weather is going to change, we need to look at the causes of such outside interference. And that means, of course, that we must look at the Sun.

To put the snow- and ice-cover effect into perspective, we can compare the percentage changes in cover during the years of the survey. Before 1971, the average cover over a year was about 33 million square kilometres. During 1971 alone, the cover increased from 33 million kilometres to 37 million square kilometres – an increase by 12 per cent in just one calendar year. And since 1971, the new average seems to have become established, with the fluctuations in snow and ice cover over each calendar year varying around 37 million square kilometres. The minimum since the increase has been 36.7 million square kilometres, and the latest maximum recorded was 37.5 million square kilometres.

Another way of looking at the situation is to compare the dates on which S.C.-autumn began in each of the 7 years. This season, the period when the extent of snow cover increases from 15 to 55 million square kilometres in the northern hemisphere, started as late as October 5th or October 10th for the years 1967-70. In 1972, however, S.C.-autumn started on September 17th, and in 1973 it started on September 20th.

One theory the climatologists are considering is that the decline in temperature over 30 years or so had reached a stage in 1971 where more extensive snow and ice cover was inevitable. Then the extension of this cover led to a change in global circulation with repercussions around the world. That may be over-simplified. But if there are changes in the amount of energy reaching the Earth from the Sun, then changes in snow and ice cover could act to magnify the effect produced.

Before we go on to look for possible causes of such changes, it is worth taking a second look at the studies of how Ethiopian droughts and the flow of the Nile can be related to past ice ages. The situation today is certainly still some way from a true ice age. But, as we saw in Chapter 3, archaeologists and geologists have now linked the occurrence of extreme drought in Ethiopia, and a southward movement of the Sahara, with the conditions of a full ice age.

When we examine recent meteorological measurements in the Sahel and Ethiopia, we find, as Winstanley pointed out, a decline in rainfall and a southward shift of the Sahara. And at just the same time, so the weather satellite pictures inform us, there is a significant increase in the snow and ice cover of the northern hemisphere. This is exactly, in miniature, the situation which the experts now believe to have existed in the last ice age.

So the different pieces of the puzzle fit together very nicely. We are indeed entering a period of at least a mini ice age. Bits of evidence as diverse as shells collected on the banks of the Nile and satellite photographs of Baffin Island all point to the same conclusion. For any one piece of evidence, it's possible to find an 'expert' who will dismiss it as of no account. But this is just the kind of situation where narrow expertise is no use; what is needed is the broader view of a non-specialist, to show how the contributions from the various experts fit together. Just the same kind of non-specialist approach is needed to sort out possible causes.

5 The Sun and the Weather

There is now no doubt that the Sun's activity affects the circulation of the Earth's atmosphere, and thus has an influence on weather and climate. The slightest hiccup of the Sun affects the Earth's atmosphere measurably; and variations of solar activity over the roughly 11-year sunspot cycle can be related to variations in weather. Therefore all solar cycles are of the greatest importance in predicting the weather. But in order to understand them it is best first to look at how specific events on the Sun can be related to specific changes in the weather on Earth.

One clue came from experiments with high-altitude rockets launched from the Kiruna test range in Sweden. These showed that the so-called 'noctilucent clouds' in the atmosphere contain relatively large quantities of heavy elements such as lanthanum, osmium and tantalum. Noctilucant clouds are extremely high in the atmosphere, at altitudes of 90 kilometres, compared with the 10 to 15 kilometres typical of normal clouds. These clouds are illuminated by the Sun long after sunset at ground level; hence their name. It had long been known that these clouds were made up of dust particles reflecting the Sun's light, but the nature of these particles was only revealed when the sounding rockets brought samples back to the ground early in the 1970s.

According to the team from the Dudley Observatory in New York, who analysed it, the dust could only have come from the Sun. Heavy metals, condensed into dust particles by some unknown process, must have been blown away from the Sun in the solar wind. The solar 'wind' is a stream of electrically charged particles – electrons and ions (atomic nuclei stripped of their electrons) – which 'blows' along the lines of force of the Sun's magnetic field. This relationship between the particles of the solar wind and magnetic field lines ties in with the location of noctilucent clouds – the polar regions – for the solar wind of charged particles is funnelled into the polar regions by the Earth's magnetic field. Once they get into the upper atmosphere, small particles such as these can act as 'seeds' on which crystals of ice and snow can grow. If these ice crystals then melt, rainfall is produced. In other words, in some circumstances and in some parts of the world, the Sun can cause rain.

There may be other ways in which solar activity and the solar wind act to affect the weather. It is by no means clear that this noctilucent cloud process is the only such mechanism – indeed, the evidence suggests that it is not.

Dr Walter Orr Roberts, an American meteorologist, has been studying links between solar activity and our weather for twenty years now, and over that time the subject which was regarded as a fringe science has grown into full respectability. He and various co-workers, have focused attention on what happens to low-pressure systems forming over Alaska just after bursts of particles from the Sun arrive at the Earth. Alaska is a good place to study such effects, because of the way solar wind particles are directed towards the polar regions by the Earth's magnetic field. No elaborate scientific apparatus is necessary to detect the arrival of particularly intense bursts; they advertise themselves, just like many other products, with displays of coloured lights – in this case, the aurora borealis. All the equipment the observers need is their eyes.

So what happens to the low-pressure systems in the Gulf of Alaska just after bright auroral displays? According to Dr Roberts and his colleagues, a few days after such disturbances, the low-pressure systems forming tend to be unusually deep and vigorous. The solar activity does not cause extra depressions to be produced, but it does seem to affect the development of the depressions which are already lining up to sweep into Alaska from the west. If proven, this is clearly an important link between solar activity and the weather. For many years, however, there was some doubt about this work. When some other groups of meteorologists and atmospheric scientists made their own calculations of the statistics involved, they could not find any evidence of such an effect.

Although Roberts continued to find fresh evidence to support his theory, conclusive evidence did not appear until 1973 and 1974. The evidence came from Europe, where the behaviour of the atmosphere is monitored at, among other places, Zugspitze, 3 kilometres above sea level in the Bavarian Alps. Since 1969, the level of certain radioactive elements in the atmosphere has been monitored at Zugspitze as an indication of the amount of interchange which goes on between the upper atmosphere (the stratosphere) and the lower atmosphere (the troposphere). The radioactive elements are produced when charged particles from the Sun — solar cosmic rays — interact with the stratosphere; their presence nearer to the ground provides a good guide to how much mixing is occurring between the stratosphere and the troposphere.

Using standard techniques, the observers record the amount of Beryllium-7 and Phosphorus-32 in the atmosphere at Zugspitze. Marked increases in the levels of these isotopes (radionuclides) mean that a mass of air originating in the stratosphere has penetrated to the observing station. In order to relate such increases to the activity of the Sun, the observers compare their occurrence with the dates of solar flares. These flares are simply bursts of activity in local regions of the Sun, and they produce bursts of solar cosmic rays. The flares can be detected by changes in the amount of X-radiation coming from the Sun, which can be measured by satellites such as Explorer 35, or by other space probes. It turns out that two or three days after a solar flare there is an increase in the amount of stratospheric air which penetrates the lower troposphere.

One obvious question arises from this work. Presumably, the amount of radionuclides present in the atmosphere is itself increased when the Sun is more active, since the flares produce more cosmic rays. It might seem that this effect could explain the phenomenon. But the measured increase in Beryllium-7 and Phosphorus-32 at Zugspitze is often between 50 and 100 per cent, and this is far more than can be explained by the small increase in cosmic rays produced by the flares. Most of this effect must be the result of extra mixing, which occurs after bursts of solar activity.

That evidence obviously lends powerful weight to Robert's ideas. If the Sun's activity disturbs the atmosphere even at the latitude of Zugspitze, it is hardly surprising that it should affect atmospheric troughs near Alaska.

In 1974, Drs Harold Stolov and Ralph Shapiro, who had previously failed to find evidence confirming the accuracy of Roberts's studies, reported that they had made a mistake. Their new statistical examination of the evidence corroborated Roberts's conclusions. This was particularly important since Stolov and Shapiro were using about the best data available: figures covering the entire period from January 1st, 1947, to December 31st, 1970. They consist of records of sea-level pressures and the altitude of the 700-millibar pressure level in the atmosphere, measured at intervals of 5° in latitude and 10° in longitude over a diamond-shaped grid covering latitudes between 20° and 70°N. These pressure measurements were compared with 'solar geomagnetic incidents' – measurable effects of the solar cosmic rays on the Earth's magnetic field – which were found in various records.

So why did the two meteorologists fail to find the link between solar activity and atmospheric pressure in their first analysis? Quite simply, they forgot to take account of the variations in atmospheric pressure over the seasons! In their more recent analysis, with seasonal changes subtracted, the geomagnetic effect is clearly shown. In their own words, there is 'firm statistical evidence' of 'a real relationship between solar geomagnetic disturbance and the subsequent behaviour of the 700 millibar height'. In winter, the effect occurs four days after the geomagnetic disturbance, and can produce an increase of 7 per cent in the mean westerly flow of the atmospheric circulation. In summer the effect is smaller, but still measurable, and occurs two days after the disturbance. This has a particularly pronounced effect on the development of low-pressure centres in the latitude belt 40°N to 60°N, which happens to coincide very closely with the latitude extent of Britain.

Proof that the Sun's activity affects the weather confirms the idea, which has been around for over a hundred years, of a connection between solar cycles and the weather. Studies of variations in the solar wind, measured from Vela and Pioneer spacecraft, provide a clue to the next step in this chain of cause and effect. It seems that the solar wind is more 'gusty' around the time of maximum sunspot activity. What this means is that the streams of particles from the Sun include more high-speed streams in a year of sunspot maximum, such as 1968. For example, in 1968 there was gusty high-speed streaming activity in the solar wind on 230 days out of the full year; but in 1971, as the Sun's activity declined, on only 73 days out of the 270 monitored by the satellites. So there is the statistical proof of what we would expect from commonsense: when the Sun is more active, producing flares and spots, the solar wind contains more high-speed streams. And these high-speed streams are very likely to affect the weather on Earth.

One of the most recent studies of how this effect works to produce an 11-year cycle of variation in the weather has been made by Dr J. W. King, of the Appleton Laboratory near Slough in England. Dr King has gathered together evidence that has been put forward over many decades to show that variations and trends in the Earth's climate may be associated with changes in the solar radiation over the sunspot cycle. Agricultural records are particularly useful since they include measurements of the length of the growing season of crops, which is defined in a very precise way. The growing season is the period of the year during which the air temperature 1.25 metres above the ground exceeds 5.6°C at the recording site. And, as Dr King points out, the length of the growing season at Eskdalemuir in Scotland, for instance, follows very closely the variations of the solar cycle of activity, as revealed by counts of the number of sunspots each year (see Fig. 9).

But there are many other ways of studying climatic fluctuations, some more subtle than others. One such indicator relates solar activity to sport on Earth – in particular, to cricket. According to *Wisden,* the cricketer's Bible, there have been only 28 years when individual cricketers have scored more than 3,000 runs in a season, and that target has only been achieved by 2 or more batsmen together the same season in 5 years. Sixteen of the 28 occasions, and all 5 of the years in which the target was reached by more than one cricketer, correspond to years either of sunspot minimum or sunspot maximum. In addition, 13 of the 15 occasions on which one batsman has scored 13 or more separate hundreds in a season were within a year of sunspot maximum or minimum. Devotees of the English game will realize just how strong an indicator of the link between solar activity and the occurrence of fine summers this provides; however good a batsman is, he cannot score a lot of runs in a wet summer.

Another piece of direct evidence linking sunspots and the weather comes from records of the occurrence of storms and lightning in Britain. Though it had been suspected before, this phenomenon was only publicized in 1974; Dr King's work encouraged Dr M. F. Stringfellow, of the U.K. Electricity Council Research Centre, to publish his findings. Once again, the evidence is best presented as a pair of graphs (see Fig. 10). The 'annual lightning incidence' is worked out from a rule of thumb which is familiar to meteorologists and to electrical engineers who are responsible for maintaining power lines, which are particularly prone to lightning strikes. Many years of observation have shown that the mean incidence of lightning (which is expressed as the number of flashes occurring in a given area each year) can be found by taking the mean number of thunderstorm days noted by meteorological observers, and raising this number to the power 1.9. Since this is only a rough guide, the square of the number of thunderstorm days can be used (the power of 2 instead of 1.9). This was how Dr Stringfellow worked out the numbers used in the top graph of Figure 10 with the meteorological data coming from 40 meteorological observers, representative of the whole of Britain, and covering a period from 1930 to 1970.

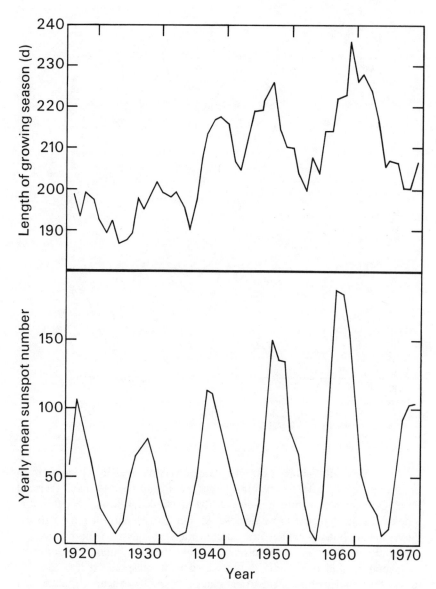

Fig. 9 Top, the length of the 'growing season' at Eskdalemuir in Scotland (55° N, 3° W); bottom, yearly mean sunspot numbers. (Dr J. W. King, *Nature*)

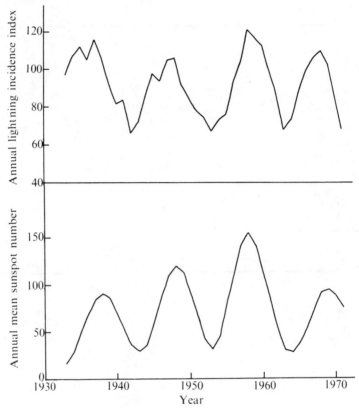

Fig. 10 top, annual lightning index (calculated as described in text); bottom, yearly mean sunspot number. (Dr M. F. Stringfellow, *Nature*)

There are many similar examples – the levels of lakes in Africa seem to follow the solar cycle, for example – but there is no need to detail them all here. One more case should, however, be mentioned, not only because it is one of the most recent to be discovered, but also because it relates to the monsoon in India – which first alerted meteorologists to the dangers of climatic change. It has been discovered that the distribution of southwest monsoon rainfall over India is associated with sunspot activity, and this discovery provides one of the direct clues that the large-scale circulation of the atmosphere is tied to local effects. The monsoon winds are just about as far from the poles as they could be; if even those winds are affected by the amount of solar particles being funnelled down on to the Earth's poles by its magnetic field, then it seems likely that the whole atmospheric circulation, and the weather throughout the world, is affected.

Although the monsoon rains show a regular yearly pattern, the amount of rain that falls during the monsoon in good and bad years can vary greatly. One of the worst droughts on record in the plains of India was in 1899, when rainfall was 26 per cent below normal. By contrast, 1917 was a year of very heavy rainfall, with precipitation 29 per cent above normal. So almost twice as much rain falls in the best years as in the worst. On a more local scale the effect is even more pronounced. The variability of rainfall is more than 100 per cent in some areas, and is least (no more than 20 per cent) over northeast India and the western coast, where rainfall is more reliable. Because of the great economic importance of such effects, they have been studied in detail by scientists at the Institute of Tropical Meteorology in Poona, who reported late in 1973 the discovery of a link between the changes and the Sun's activity. Their analysis reveals 'composite patterns' in the rainfall variations, and these can be related to four states of solar activity: sunspot maximum, sunspot minimum, sunspot increasing and sunspot decreasing. Although they did not mention the present serious droughts, we are in a period of low solar activity (1899 was a time of low sunspot activity and 1917 was a time of high sunspot activity). But a very important word of caution is needed before this kind of link can be used to predict climatic changes.

There is no doubt that other factors, apart from variations in solar output over the sunspot cycle, must affect our climate. These might act in the same direction as solar effects, increasing changes in the weather, or they might act in the opposite direction, so that all the different trends effectively cancelled each other out. Clearly, solar influence has been acting so as to produce drought in India and the Sahel. But the solar cycle is not the only factor; longer term trends are influencing the weather in the same way, but the addition of the solar effect may have brought about the recent sudden changes. If that is correct, things will not change so quickly in the next few years, because we have just reached a minimum in the cycle of sunspot activity, and are heading towards a peak around 1982. But if the long-term trends which have been operating now for two or three decades continue into the 1980s, then in eight or nine years' time the solar effects will again be adding to them, and there will be another period of several years in which the weather dramatically worsens. But then the change will start from conditions like those of today – and those conditions are already markedly worse than the conditions of the mid-1960s.

How long will all this continue? Well, the two parts of the puzzle must, it seems, be solved separately. Let us first examine in more detail how solar activity varies, and what are its likely effects on the weather.

The link between sunspots and the weather – in particular, the effect on the growing season – is significant even over a decade or so. We can be confident that certain crops will do better in Britain and America in 1982 than they did in 1974, just by extending the graphs in Figure 9. But, as that figure also shows to a small degree, the strengths of sunspot periods differ: some years of peak activity have many more sunspots than others. These variations are reflected in the variations of growing season length, number of thunderstorms, and other effects on the weather.

Although sunspots were known to the ancient Greeks, this knowledge was lost and only rediscovered by Galileo, at the beginning of the seventeenth century. And systematic counts of the number of sunspots seen each year only go back to about 1800. So studies of the way in which variations in the strength of the sunspot cycle affect weather and climate are restricted to only just over 250 years' worth of information. Any variations over that sort of time-scale hardly count as long term, in the history of the Earth or even in the history of mankind. But there is just enough information to highlight one fascinating correlation.

Many astronomers during the past hundred years or so have puzzled over the nature of the sunspot cycle. The puzzle is made more difficult because although the cycle is generally recognized as an 11-year variation, individual cycles can be as short as 8 years or as long as 13 years (in rare cases, even longer). With the variations in strength – that is the differences between the numbers of sunspots seen at different years of solar maximum – occurring as well, it is clear that the process generating sunspots is a complicated one. Today, sunspots are recognized as being just the most visible symptom of a fundamental cycle affecting the Sun's activity. Magnetic fields, flare activity, the solar wind and other properties of the Sun all vary in the same way.

The cause of the sunspot cycle is still largely a mystery, but one influence has been pinned down. It did not take astronomers long to think of the idea that the cycle might be linked to Jupiter, because the period of Jupiter's orbit around the Sun is just over 11 years. Jupiter is the planet most likely to affect the Sun, since it is by far the biggest planet of our Solar System. The sunspot cycle does not exactly coincide with Jupiter's orbital movements, but the difference can be explained by taking the effects of the other planets into account.

Throughout the twentieth century various people have shown that the Earth, Mercury, Mars and Venus all seem to produce an effect on the solar cycle. For example, when the Earth and Venus are aligned on the same side of the Sun the number of sunspots is different from the occasions when the Earth, Sun and Venus mark the corners of a right-angled triangle. Though this might seem like astrology, there is a very good reason why planetary alignments should affect the Sun: just as the Sun and Moon raise tides in the oceans and atmospheres of Earth (and even in the solid earth itself) so the gravity of the planets raises tides on the surface of the Sun. And the height of the total tide depends on the relative alignments of the planets.

This tidal effect can be calculated quite accurately. Professor K. D. Wood, of the University of Colorado, has compared the tide raised on the Sun by the influence of Venus, the Earth, Mars and Jupiter with the variations of the sunspot cycle since 1800. The result, shown in Figure 11, speaks for itself. The year of sunspot maximum or minimum coincides closely with the year of tidal maximum or minimum, and there is a less precise link between the height of the tide at each maximum and the strength of the corresponding solar cycle.

Professor Wood is already able to use this to predict the year of the next solar maximum as 1982 – it seems that the present sunspot cycle is longer than average. This prediction was made early in the 1970s, only just after the peak of the most recent sunspot cycle. By the beginning of 1975, observations of sunspots had shown that the present solar cycle is developing exactly in line with Professor Wood's prediction. Like other astronomers, Professor Wood has also noted that there seems to be a repeated pattern in the variations of the strength of the sunspot periods. It looks as if the pattern repeats every 180 years or so – but since we don't have much more than 180 years of measurements, it is difficult to know if this is a real periodic variation or just a coincidence in the few hundred years of data we have.

When such a suspected periodicity crops up in observations but cannot be proved, the obvious thing to do is to try to find a physical reason why such a period should exist. A physical cause would give more credence to the idea of such a periodic variation. If we calculate the influence of all the planets, including the other planets out beyond Jupiter (whose tidal effect is small, because of their great distance from the Sun), we find a periodicity associated with their movements which varies in just the way we are looking for. This is the movement of the centre of mass of the whole Solar System.

Fig. 11 Professor Wood's evidence for a link between sunspot activity and the tidal influence of the planets on the Sun since 1800. The dashed line shows variations in sunspot number; the solid line shows the normalized value of the tide raised on the Sun by the 'tidal planets'. Sunspot cycles are numbered according to a convention by which the cycle peaking between 1800 and 1810 becomes no. 5 (Professor K. D. Wood, *Nature*)

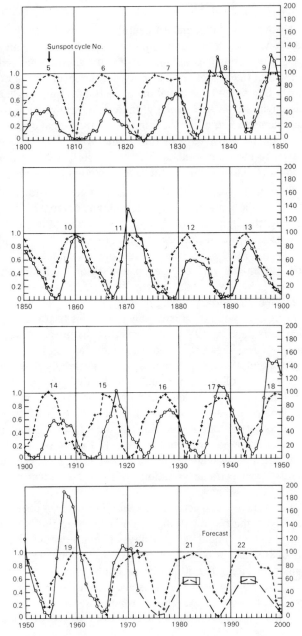

Because the Sun is so much more massive than all the planets put together, this centre of mass generally lies within the Sun itself. In a similar way, the centre of mass of the Earth-Moon system lies within the Earth. But the Solar System is much more complicated than the simple Earth-Moon system, and as the planets move in their separate orbits the centre of mass, which depends on their average effects, moves in a looping path in and around the Sun. This path spreads over about the same distance as the diameter of the Sun, and occasionally the centre of mass actually moves just outside the solar surface. Could this stirring effect be related to sunspot activity? It seems possible. The combined effects of the planets produce patterns in the movement of the centre of mass with periods of 11, 22, 30 and 80 years, all important periods for solar variations. And we can understand why there might be a physical link by considering not how the centre of mass moves relative to the Sun, but how the Sun moves relative to the centre of mass. In effect, what is happening is that the combined gravitational pull of the planets is swinging the Sun about in a peculiar looping orbit around the centre of mass − and that could well produce centrifugal forces which might influence the internal workings of the Sun, and the visible spots.

The most unusual alignment of the planets occurs when they are all lined up on the same side of the Sun − and that happens every 179 years. The circumstantial evidence is impressive. If the sunspot cycle has a long term periodicity of 180 years (which is close enough to 179 years to make no difference) then it is very tempting to relate this to the recurring grand alignment of the planets. The next alignment of this kind is due in the 1980s; but we may already have seen the turning point of the related cycle of sunspot activity and climatic change. Other alignments − for instance, with all the planets except Jupiter on one side of the Sun and Jupiter on the other − might be more significant in terms of sunspot activity. All such alignments repeat every 179 years.

But tidal theory of sunspots has come under attack, notably in a study carried out by Don Anderson and Emile Okal, of the California Institute of Technology. They have been encouraged by studies like that of Wood to take this kind of calculation a step further, and to work out the variations in the tides on the Sun that are produced by Mercury, Venus, Earth and Jupiter taking into account the complete details of the variations of these planets' orbits as they move around the Sun.

This requires rather more sophisticated calculations than those used by Professor Wood, since the more recent study allows for such effects as the inclination of each planet's orbit relative to the equator of the Sun, and how this varies from time to time. With this extra sophistication and (unlike Wood's calculations) with the Mercury influence included, the CalTech. workers can find no evidence of a link between the height of the tide raised at any time and the sunspot number at that time. According to Anderson and Okal, the effect noticed by Wood was just a coincidence between two very similar periodic cycles (averaging 11.86 years for tides and 11.05 years for sunspots) which happened to overlap for a large part of the period covered by Wood's study.

This work rather begs the question of the movements of the centre of mass of the Solar System over the 179-year cycle. Furthermore, it's certainly not yet true that all astronomers are convinced by the CalTech. study, just as not all are convinced by the tidal theory of sunspots. Unfortunately, this is still an area of science where a lot of loose ends remain to be tied up. But whatever the theorists may say about the *cause* of the link between sunspots and planets, there is still a lot of evidence that certain alignments of some planets do indeed seem to be responsible for particular bursts of solar activity.

One of the most intriguing examples of this link between planets and solar activity is provided by looking at the planetary alignments at the times of two great solar outbursts, one of which took place in 1960 and the other in 1972. Because bursts of solar activity affect the Earth's magnetosphere they can disrupt radio communications around the globe, radio engineers are among the keenest students of sunspot cycles. One such engineer, Dr J. H. Nelson, produced the chart shown in Figure 12(a) *before* November 12th, 1960, as a prediction for N A S A that this particular alignment would cause an outburst of solar activity. That outburst duly occurred. More than a decade later, while I was involved in a study of the effects of a still greater solar outburst (in fact, the largest burst of solar activity ever recorded) I came across Dr Nelson's 1960 prediction and out of curiosity asked him what the alignments of the same planets looked like on August 4th, 1972, when the outburst I was investigating had occurred. His answer came back in the form of Figure 12(b), together with a comment that this particular alignment of the same three planets to form one right-angle triangle and one 120° triangle with the Sun occurred only on these two particular occasions during the entire 12-year period from November 1960 to the end of 1972.

Even with evidence such as that of Figure 12(a) and (b) the astronomical controversy about mechanisms continues. For the purpose of the study of climate on Earth, however, this need not concern us; all we need to know is

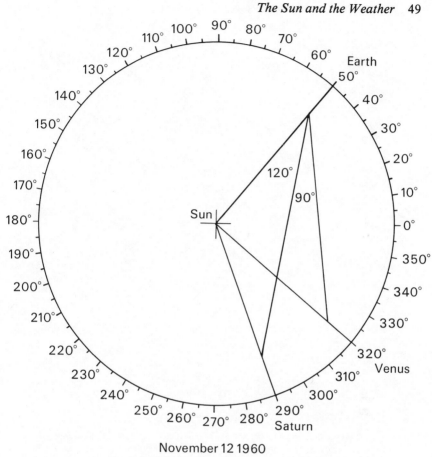

Fig. 12a Alignment of Saturn, Earth and Venus at the time of the solar storm which occurred in mid-November 1960. (Dr J. H. Nelson)

that there are real, predictable changes in the level of solar activity. As it happens, the records seem to show that 1959, rather than 1982, marked a turning point in solar activity. After several successive increases in the strength of the sunspot periods during the first half of this century, there has been a decline since 1959. And sophisticated statistical analysis of the sunspot number variations bears this out, making it possible to predict future changes with even more confidence.

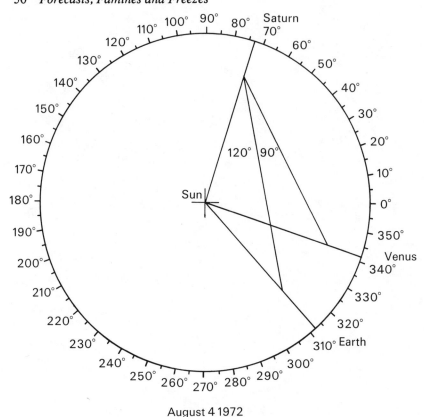

August 4 1972

Fig. 12b Alignment of the same three planets at the time of the solar storm which occurred early in August 1972. Could these alignments be coincidental? Two such occurrences cannot be taken as *proof* that the alignment in some way causes the solar activity, but these were the only two occasions in the 12 years, 1960-72, that this particular combination of Earth, Venus, Saturn and the Sun to produce one 90° and one 120° triangle occurred. But whatever causes the variations in solar activity, there is no doubt that these variations in turn affect the weather on Earth. (Dr J. H. Nelson)

This particular analysis was carried out by Dr Theodore Cohen and Dr Paul Lintz, of Teledyne Geotech in Alexandria, Virginia. They have used the 'number crunching' approach – a powerful high-speed electronic computer is programmed to drag out any evidence of periodic variations from a set of numbers. In this case, the numbers are the records of sunspot activity, and the name of the technique used for this problem was 'maximum entropy spectral analysis', or M E S A .

As we have seen in Figure 9, in some years of peak activity the Sun may produce 200 sunspots, while in other peak years the number may be as low as 50. So there is plenty of variation to analyse. The result of this particular analysis confirms that there is a 179-year variation. But it also throws up some other periodic changes, and Drs Cohen and Lintz believe that it is the interaction of the shorter term effects that produces the 179-year period. Their argument is that two of the periodic variations they have found (with periods of 11.2 years and 9.8 years) can produce a 'beat' effect which has a period of 179 years. This beat effect is essentially the same as the process which produces an audible beat when two very close, but not identical, musical notes are struck together.

The snag about the 'number crunching' method is that it is completely divorced from physical reality: nowhere does the problem of what causes the variations enter into the calculation. For example, the Teledyne team also finds a strong periodicity at 89.6 years; it doesn't take a mathematical genius to notice that this is exactly half of the fundamental 179-year period between grand alignments of the planets, nor to notice that 89.6 is just eight times the chief period found by the M E S A technique, which is 11.2 years. The power of the number-crunching approach is also its weakness. For, if we are prepared to accept the results without worrying about the physical causes of the periodic variations, we can use them to predict future solar activity regardless of 'the reason why'. What the prediction tells us is that the present solar cycle is indeed on the long side, and will peak in 1982 at the relatively low maximum of 50. Peak sunspot numbers above 100 cannot be expected again until about 2015.

Accepting the recently discovered link between solar activity and the weather without reservation, we should expect the growing season to be shorter, on average, in the next 40 years than it has been in the past 40 years (quite apart from any longer term trends in the climate). This will seriously affect agriculture. The comparative 'failure' of the 1974 harvest might well be looked back on with envy a few years from now. If these solar effects are related to the increased ice cover discussed in Chapter 4, we can also expect more ice in the near future. This will affect the fishing industry, which would otherwise seem the obvious method of making up the deficit we are going to suffer in land-based food production. The effects of declining temperature and rainfall changes on industry have already been outlined in Chapter 1. If the long-term trends, as well as the sunspot variations, are running against us for the next 40 years, then we can expect something like a mini ice age − a return to the conditions which have, in fact, been normal in Europe and America for most of the past millennium, with the first half of the twentieth century standing out as an unusually warm period, a little climatic optimum. And, according to some theories, full ice ages may be caused by longer term changes in the output of the Sun (see p. 60).

6 Long-term Climatic Trends

Climatic trends have only been studied scientifically for a fairly short period of time. But, by using meteorological information from many kinds of historical records, it is possible to build up a good picture of climatic change in historical times. As we have already seen, the records of the twentieth century show a warming up, until about 1950, an apparent emergence from the cold conditions that had dominated for hundreds of years before. But since 1950 there has been a general cooling, back towards the normal state which has persisted for a thousand years. So it seems that the first half of this century − just the period when meteorological standards were being established − was in fact a very unusual warm spell. A better standard of comparison can be found by looking at the historical records.

It was only in the middle of the seventeenth century that the barometer and thermometer were first invented, so we do not have any really accurate measurements of temperature and pressure going back further than 300 years. Because it took a while for the use of these instruments to be adopted, widespread measurements, which provide the data for working out world weather charts, only go back for a couple of centuries. But even they provide a valuable pool of information.

The early observers were chiefly amateurs, often doctors and clergymen, such as William Derham, the vicar of Upminster in Essex, who kept daily observations of the meteorological elements for many years, starting in 1697. One amateur meteorologist who must have been quite a character was E. A. Holyoke, a physician at Salem, Massachusetts. He kept a record of observations from 1754 until shortly before he died in 1829 at the age of 100. Modern meteorologists regard his measurements as far superior to those of an official observatory in the same area − but Holyoke himself was such a perfectionist that at one time he asked that the records he kept after his 92nd birthday should not be published, since they did not come up to the standards he maintained when younger!

Official records are also important, of course, and ships' logs are a useful source of information. Many of these bits and pieces are collected in the *World Weather Records,* published by the Smithsonian in Washington. Using such information, meteorologists can construct weather charts covering Europe, most of the Atlantic Ocean, and the Eastern U.S. for the past two centuries. These are not as detailed as modern daily weather charts, but indicate the broad trends.

Such work provides a remarkable example of painstaking scientific study being rewarded many years later. Only now are the studies of many individuals working in the eighteenth century reaching fruition, as present-day meteorologists go through the long job of constructing weather charts from their observations.

One thing emerges clearly from even the scanty information available for the earliest period covered by these measurements. The atmospheric circulation was weaker then; that is, warm, damp Atlantic air could not penetrate so well into Britain and the western part of Europe. That ties in with, and explains, historical reports of more extreme conditions in Britain two or three centuries ago, with frequent severe frosts. The Thames, for example, froze over several times in that period, and ice cover in the Arctic Ocean was more extensive. This is just the same situation that Winstanley and others believe to be repeating today. Once again, the atmospheric circulation is becoming weak. Ice cover in the Arctic is, as we saw in Chapter 4, increasing. And the change in circulation is causing a cooling combined with a shift in rainfall belts.

For a picture of the weather of still earlier times, more than three centuries ago, climatologists turn to agricultural records. Years of drought, extreme frosts and floods are all documented in western Europe at least back to the year A.D. 1000. Other historical records can help: the epic voyages of the Vikings across the Atlantic to Iceland, Greenland and North America could only have been possible in a fairly mild climate, and these voyages took place over a period of a hundred years roughly a thousand years ago. Since then, the climate has been more severe; cold destroyed the Greenland colony, and agricultural records show a grim millennium, with little relief from what is now called the 'little ice age'.

The ingenuity of meteorologists in finding records of old climates is shown in particular by their use of the dates of the wine harvest. In wine-growing countries such as France, the day on which the wine harvest officially begins has long been marked by festivals, and recorded in official documents. The grapes mature more quickly in good summers, so these dates can be used to study climatic change. These measurements only show relative change – we can say that one year was 'better' than another – but the wine harvest dates do not give an exact indication of temperature and rainfall. In a way, this is a bit like the use of varves or tree rings. And in fact for certain periods of a few years the varve and tree-ring patterns exactly parallel the wine harvest dates. All three methods combine to confirm the trend of climate.

It would be possible to write a book about these measurements and indicators alone – indeed, the French historian Emmanuel le Roy Ladurie has done just that.* We can gain a powerful insight into the value of historical records by looking outside Europe – across to the other side of the world, where the oldest man's civilizations provides the only complete run of historical meteorological information covering the past 5,000 years.

The Chinese records are valuable both because they cover such a long time and because they come from a completely different part of the globe; the more recent Chinese records (for the past millennium) are complementary to the European records. This vast pool of information about historical changes in climate is being studied by many Chinese meteorologists and historians, one of whom, Chu Ko-chen, reported their preliminary work in the Chinese scientific journal *Scientia Sinica* in the middle of 1973. It is a tradition in modern China that even erudite scientific papers are sprinkled with quotations from the thoughts of Chairman Mao. But some of these are singularly appropriate for this long-term study of ancient records; in particular, Chu Ko-chen refers to Mao Tse-tung's policy of 'making the past serve the present'.

According to Chu, the historical records fall naturally into four sections:

The archeological period, from 3000 B.C. to 1100 B.C., when the only 'written' records were those carved on the so-called 'oracle bones'.
The documental period, from 1100 B.C. to A.D. 1400. For this period, there are written records of a more conventional kind, but there are no detailed reports covering the different regions of China.
The gazetteer period, from A.D. 1400 to 1900. During this 500-year long period, many districts kept their own official records of history and geography, and these contain a wealth of climatic information.

* *Times of Feast, Times of Famine* (Allen & Unwin, London, 1973).

The instrumental period, from 1900 onwards. This is the modern period, with proper detailed meteorological records covering all parts of China.

Even for the earliest of these periods important climatic evidence is available. The oracle bones, which contain many inscriptions praying for rains, provide an important clue; evidence of farming activity, and farming methods, provides another. It seems that since the time of the Yangshao culture, 5,000 years ago, the northern limit of bamboo has moved south by 3°. This corresponds to a decline in temperature along the lower reaches of the Yellow and Yangtze rivers of about 5°C (on average) for January and some 2°C for the mean annual temperature. Yangshao and Yin-Hsu times, at the earliest period of the 5,000 years studied, seem to have been a climatic optimum in China, when rich sub-tropical fauna and flora flourished in places like Sian and Anyang. But 'as to the existence of elephants in the northern part of China during the neolithic age, it is still a moot question,' says Chu Ko-chen.

In some ways, the analysis of evidence from the documental period is the most fascinating. Here, there is no lack of data – the problem lies in interpretation to giva self-consistent picture. The keeping of official records began with the founding of the Chou Dynasty (1066 B.C. – 256 B.C.), with first inscriptions in bronze and later written records in bamboo books. But the official historians were not the only people to leave records, and poets in particular have left descriptions of birds and flowers from which evidence of climatic change can be deduced. Certainly, at the beginning of the Chou Dynasty the climate was warm enough to grow bamboo extensively in the Yellow River Valley. But the climate soon deteriorated, and the Han River, a tributary of the Yangtze, froze in 903 and 897 B.C. Great droughts followed the freezing, but conditions in the mid-Chou period were rather better. Plum trees, much loved by the Chinese, became common throughout the country, and from about the ninth century B.C. to the time of Confucius (557-479 B.C.) the Yellow River valley was slightly warmer than it is today. The next cold period began about 200 B.C. and in the space of a century there were six very severe winters, with frost and snow occurring in late spring months, and many people froze to death. From the first century B.C. to the first century A.D. the climate again recovered, but between A.D. 155 and 220 attempts to grow oranges in the royal garden at Loyang 'failed dismally', even though oranges had been grown in the royal gardens during the Han Dynasty some 300 years earlier.

The astronomer-poet Chang Heng (A.D. 78-139) had written an ode in which he referred to plentiful orange groves near the 'South Capital' – but in A.D. 225 naval manoeuvres at Kuangling had to be suspended when the Huai River froze, for the first time on record. So the pattern of warm and cold periods continues; it is plausible to argue, for example, that during the eighth and ninth centuries A.D. temperature and rainfall belts in the eastern part of China were a little north of their present-day positions, and Chu progresses through the maze of documental evidence up the twelfth century. Then, there was a much more dramatic change in the climate.

In the past few years, climatologists (and even some non-experts) have come to be familiar with the idea of a 'little ice age' which affected Europe and Britain and which may have been at least partly responsible for the loss of Viking colonies in North America and Greenland. But in China, it seems, this little ice age was pronounced in the twelfth century, considerably earlier than in the west. Japanese records confirm this; from the ninth century onwards, nobles celebrated the blooming of Japanese cherry trees (a sensitive climatic indicator) and appropriate records were kept until the nineteenth century. Climatic changes in China and Japan paralleled each other during that millennium. The long cold period lasted effectively from the tenth to the fourteenth centuries in China, with the severe cold of the twelfth century corresponding most closely with the western little ice age. Development and movement of this cold period began in European Russia around 1350, and in central Europe between 1429 and 1465; in England it was during the sixteenth century that conditions became so serious that they led to the Elizabethan legislation on poverty. Professor Hubert Lamb has already pointed out that 'the greatest incidence of anomalies moved westward across Europe during the climatic deterioration from 1300 to 1600 and returned to the East during the recovery from 1700 onwards.' Extending this pattern of climatic movement to China and Japan provides one of the most powerful tools yet for climatic research and prediction of future trends.

Records from the gazetteer period provide, among other data, further evidence of this east-west movement of climatic change. Freezing of lakes indicates unusually severe winters, and although this kind of evidence from China and Japan shows that the seventeenth century had the most severe winters in both countries, in Japan the period of severe freezing started and ended 25 years before the corresponding period in China. The evidence from the instrumental period seems almost mundane compared with the importance of the historical data, but this shows a pattern of upward and downward swings in temperature, by between 0.5 and 1°C, which is obviously of great significance in predicting future trends, particularly to the farmers. This kind of swing, with a period of rather more than 40 years, has been going on for thousands of years, although sometimes obscured by more dramatic changes.

In summary, beginning 5,000 years ago: for the first 2,000 years China was about 2°C warmer than today; then a series of up and down swings with minima at about 100 B.C., A.D. 1200 and A.D. 1700 and amplitude 1°-2°C can be traced from the more complete data available. Within each 400-800 year period, smaller cycles of 50-100 years in length can be traced, covering 0.5°-1.0°C in their variation; for both kinds of cycle, the coldest period begins at the Pacific coast of Asia, moving west as a wave right through to the Atlantic coast of Africa and Europe.

So Europe, following much the same pattern of cyclic variations, usually lagged behind China. Chu argues that this is because both climates are controlled by the Siberian high-pressure system, and that fluctuations in this high affect the whole of Eurasia in a consistent pattern. As we shall see in the second part of this book, that idea ties in with some theories that have been developed by meteorologists in the West. But ideas of this kind can only be tested by analysing long-term patterns of climatic change; where they have no long-term historical records (in the Americas, for example) climatologists must turn to tree rings (as we saw in Chapter 3) and to geological evidence to find patterns to study.

The best trees for this kind of work are the bristle-cone pines of the southwestern United States. These live to a great age, and they grow in an isolated part of the world. Because they are not interfered with by man, dead wood has remained untouched for hundreds and even thousands of years. The dead wood can be fitted into the tree-ring record by comparing its overlap with living wood. If a tree now living has been growing for, say, 500 years, and a dead log from nearby was alive between 800 and 300 years ago, then the tree-ring 'fingerprint' of the outer 200 years of the dead wood will match the inner 200 years' worth of growth in the living wood. Still older wood can be dated by comparing its outer rings with the inner rings of the previously dated log, and so on. In fact, a record covering more than 8,000 years has been built up in this way from the living and dead bristle-cones of the White Mountains in the U.S. This particular tree-ring 'calendar' is mainly used for dating wood samples, not for climatic studies. But two other studies have been used for climatic work.

One of these is an 1100-year long 'calendar' built up entirely from living trees; the other built up by overlapping tree rings, is a stretch of 5,405 years, beginning in 3435 B.C., and running right through to 1970. The 5,405 year record provides an indication of variations in the temperature of the warm season (April to October) in the White Mountains. Very roughly, the pattern revealed is as follows. A short cool spell almost 5,000 years ago was followed by a period slightly warmer than the average which lasted for 2,200 years starting around 3,500 B.C. From about 1300 B.C. until 200 B.C. temperatures were generally below average, then slightly warmer again until about A.D. 1000, followed by a short, sharp 'warm spell' centred around A.D. 1200 and lasting for a couple of hundred years. Finally, from about A.D. 1400 to A.D. 1900 there is evidence of the little ice age which also affected Europe and China, with a sudden rise in temperature over the first fifty years of this century. And the total range in temperature covered by all these fluctuations? Just about 2°C!

Again, this kind of information ties in with reconstructions of atmospheric circulation patterns which suggest that a more vigorous type of circulation is related to conditions that we regard as more clement. The geological studies provide a whole family of techniques in themselves.

One technique depends on measuring the proportion of a heavy isotope of oxygen (oxygen-18) in the ice. Most of the oxygen in the atmosphere and the oceans of the world (and in rocks and living things) is the isotope oxygen-16; oxygen-18 is no different chemically, but is slightly more massive than oxygen-16 since it contains two more neutrons in its nucleus. The relative amount of the rarer form of oxygen present at any time depends on the average temperature. When snow forms and falls, the water it is made of carries oxygen isotopes in a ratio which depends on the temperature. When the snow is compressed into ice, then covered over the years by succeeding annual snowfalls, which in turn are compressed into ice, we have an ice sheet which contains within it, layer by layer, a continuous record of how the earth's temperature has varied.

Studies of changes in sea level around the world (which can be related to the amount of ice in the polar caps), measurements of the amounts of different isotopes of oxygen in the shells of marine animals which died thousands of years ago, as well as studies of the isotopes found in cores drilled from glaciers, all provide useful evidence. One ice core, 4/5ths of a mile long and 4″ thick, was drilled from the Greenland ice cap. The ice cap is continually eroded from below and replenished by fresh snowfalls from the top, with the result that each layer of snow gradually works its way down from top to bottom, a journey that takes about 110,000 years. The core pierced through the ice cap thus contains a continuous record: it reveals the pattern of 100,000 years of climatic change, recorded with a broader brush than the information from tree rings or historical records, but no less valuable. One of the most intriguing discoveries in this record is an apparent 13,000-year cycle associated with ice ages. The warm spell from A.D. 1150, when the Vikings made their epic voyages, the little ice age some 300 years ago, and other changes known from other sources are also recorded in the ice core, and help establish the accuracy of the isotope 'climatic calendar'. In the southern hemisphere, a core more than a mile long has been obtained from the Antarctic ice cap, and this too shows very long-term patterns of change.

According to that ice record, a long-term decline in temperature set in about 75,000 years ago, and reached its coldest point some 17,000 years ago. A rapid improvement in temperature then produced conditions like those of today about 10,000 years ago. This general pattern underlies shorter term variations of temperature – notably minor warmings – which occurred about 25,000, 31,000 and 39,000 years ago.

Back in 1970, scientists at Copenhagen University used the evidence from the Greenland core to make a prediction of rather more immediate importance. They found a regular pattern in which warm spells seem to be separated by intervals of about 180 years (reminiscent of the pattern of sunspot variations linked to planetary alignments, as discussed in Chapter 5) and exceptionally cool periods, sometimes separated by intervals of 80 years, seem to be related to abnormally long solar cycles. Half a decade ago, that evidence already pointed to a cooling down during the 1970s and 1980s. Sunspot activity during these decades was predicted to be low, and the prediction was also that the solar cycle now in progress (covering 1969-82) would be unusually long. Up to 1975, observations of both these aspects of sunspot variations bear out the prediction, and the further prediction that these solar effects would be associated with the beginning of a cool period also seems to be being substantially borne out. Other studies show definitely that warm spells like that of the early part of this century, are associated with periods of increased sunspot activity, and similar conditions occurred at times of strong solar activity around 1350, 1530 and 1730. The ice ages themselves may be caused by more extreme variations in the output from the Sun.

Over the past few years, some astronomers have come to view the Sun as a rather unusual star. They have been backed up by the failure of experiments designed to detect the particles called neutrinos which ought to be created by the nuclear reactions going on in the Sun. Neutrinos are notoriously elusive particles. Although nuclear theory requires the presence of neutrinos to explain how elementary particles interact, and although they have been detected in experiments on the Earth, neutrinos in the flux of cosmic rays which arrives on Earth from space have proved difficult to detect.

It is rather curious, because according to the most widely accepted theories of the internal workings of stars, our Sun should be producing a great number of neutrinos in nuclear reactions going on deep in its interior. Once formed, neutrinos are extremely reluctant to interact with other elementary particles, unless the conditions are just right. So most of the neutrinos from the solar interior should pass through the Sun's outer layers and across interplanetary space, travelling at the speed of light to arrive near the Earth a few minutes after they are formed. This very reluctance to interact with other particles makes neutrinos difficult to detect, even in laboratory tests, but experimenters have been searching for evidence of solar neutrinos since the 1950s, with a striking lack of success.

In 1972, Professor William Fowler, of the California Institute of Technology, suggested what might seem the obvious answer to this problem, but one which had not been formally proposed before: that, after all, very few neutrinos must be being made in the Sun at present. The suggestion put the cat among the pigeons. For it implies that either nuclear physicists are wrong, and that the nuclear reactions going on in the Sun produce fewer neutrinos than expected, or that the stellar physicists are wrong, and that those reactions are not going on in the Sun at all. Of the two alternatives, the second is possibly the less unpalatable.

Why have astronomers believed for decades in the need for nuclear reactions in the Sun's interior? The answer is simply that a star can exist for the thousands of millions of years indicated by geological evidence of the age of the Earth and Solar System only if nuclear energy is the driving force behind its extravagantly wasteful outpouring of light and heat into space. Without nuclear energy, a star like the Sun would collapse and die in a few million years, much less than the length of time for which life has existed on Earth.

Seemingly, there is no solution to the dilemma. But Professor Fowler does not stop there. He suggests that the lack of nuclear fusion in the Sun is a temporary phenomenon, which has lasted for only a few million years. That is but a brief span in the lifetime of the Sun, and soon fusion processes ('nuclear burning') will begin again.

What makes these ideas worthy of serious consideration is that conditions on Earth today are very different from those of 30 million years ago, and the overall climate of the Earth has undergone changes in the geological past which are known to have lasted for a few millions or tens of millions of years. Can it be, Professor Fowler argues, that the nuclear burning in the Sun periodically dies out, producing changes in the Sun's brightness which result in ice ages and other dramatic changes on Earth? Since Professor Fowler opened the way for such speculation, several other astronomers have come up with more detailed explanations of how the Sun might have gone 'off the boil' for the past few thousand (perhaps even 100,000) years. The details are not important here, but the ideas are discussed seriously at sober scientific meetings, and published in scientific journals.

We have certainly come a long way in our look at long-term climatic trends – from bristle-cone pines and Chinese oracle bones to the heart of the Sun itself. It is time surely to get back to solid earth again. So far, we have seen how even small climatic change can affect man's activities in dramatic and disastrous ways, and we have seen just how common such changes have been over hundreds of thousands of years. Obviously, we are in for further changes, both on the grand scale (which is only of academic interest) and in the immediate future. What exactly is being done to predict such changes and to prepare our modern civilization to cope with them?

7 Weather and the Earth's Magnetic Field

The Earth's magnetic field is far from constant. The magnetic poles – and the whole magnetic field – move slowly relative to the solid Earth. At present this drift is from east to west. The field can also change in strength; these changes occur either gradually or quite suddenly, when measured against the timescale of most geological processes. They must affect conditions on the Earth in many ways, and according to the latest ideas, still in the early stages of development, there is a very close link between magnetic change and climatic change. This discovery, together with a growing understanding of the Earth's magnetism, provides one intriguing prospect for predicting climatic change.

The most extreme kind of magnetic effect is a complete reversal of the Earth's field, with north and south poles literally swopping over. It seems from the geological evidence that a complete reversal of this kind can occur in less than 5,000 years; on some occasions, the change seems to have happened within less than 2,000 years. When this happens, as it has done several times, sediments being laid down at the time of the reversal are affected, so the rocks formed out of these sediments contain a magnetic record which geologists today can use to trace variations in the Earth's magnetic field back over millions of years. This magnetic record is produced because magnetic particles in rocks are aligned by the Earth's magnetic field as they are laid down. The resulting solid rock carries the characteristic signature of the Earth's field at the time it was laid down, and later changes in the field cannot affect the rock once it has solidified. When geologists study cores drilled from rocks on land and from the sea-bed, they find that successive layers of the rock, laid down longer and longer ago, are magnetized in quite different directions. Small changes in direction and changes in strength, as well as complete reversals of the field, can all be traced through this 'fossil magnetism'; and they can be accurately dated provided the ages of the rock layers are fixed by other geological evidence.

Changes in the Earth's magnetic field over the past 1,200,000 years have been mapped in detail in this way, notably by a team of scientists working at the Lamont–Doherty Geological Observatory in New York. The members of that team have been able to compare the magnetic changes with climatic changes, mainly by measuring the variation in the amount of tiny creatures (planktonic formanifera) called *Globorotalia menardii* found in the sediments. It seems reasonable to argue that more of these creatures are around in warmer weather, so when the climate is good more of their shells will be found in the sediments. On that basis, periods of high magnetic intensity are associated with colder climatic conditions. But it remains something of a puzzle just why this should be so.

One idea is that the shielding effect of the Earth's magnetic field, which protects the Earth from cosmic rays, plays a part in determining the overall atmospheric circulation and the global temperature balance. This idea has not yet been worked out in detail, but it certainly looks promising, since we already know that changes in the output of cosmic particles from the Sun do affect the weather in just this way – more particles being associated with warmer weather.

The more dramatic magnetic changes also have interesting implications for the very existence of life on Earth. During complete reversals of the Earth's field, the magnetism must, for a short time at least, die away to nothing. And that would allow a fierce torrent of cosmic rays to reach the Earth's surface. Such a flood of radiation would destroy animal life. Plants are more resistant to radiation, and marine creatures would be shielded by the sea; but land animals could be wiped out in such circumstances. Might that explain the sudden disappearance of the dinosaurs, after they had dominated the Earth for millions of years? And could rapid mutation of smaller animals during a period of stronger radiation explain the rapid early evolution of man's ancestors? that is certainly a possibility; another recent idea is that it might be the climatic changes associated with magnetic reversals that have produced these dramatic evolutionary effects.

The period of reduced field intensity during a magnetic reversal lasts for between 1,000 and 10,000 years. Drs C. G. A. Harrison and J. M. Prospero, of the University of Miami, have investigated how the resulting influx of particles into the atmosphere could affect cloud formation. These are the kind of particles which today are associated with production of noctilucent clouds high in the atmosphere around the poles (see Chapter 5).

With a period of weaker magnetism, the particles might penetrate into the atmosphere at lower latitudes and produce still more clouds. Harrison and Prospero argue that the increased cloud cover at low latitudes would affect the climate. More clouds would reflect away more of the Sun's heat, cooling the Earth. Perhaps the cooling would be dramatic enough to wipe out whole species of animals not able to adapt to these different conditions. This might help to explain the fate of the dinosaurs, but what we want to know is how changes in magnetism and the weather are related over periods of decades, not thousands of years. And once again it is work at the Lamont–Doherty Geological Observatory, this time by Dr Goesta Wollin that provides a clue.

These studies – reported only as recently as March 1973 – show how the close relationship between the Earth's magnetic field and climate has operated from 1925 to 1970. Changes in the Earth's magnetic field are monitored regularly at many observatories today, and in this study information from more than 200 observatories was gathered and compared with the annual and ten-year averages of weather conditions which are published by the United States Weather Bureau. Like the magnetic observations, the weather information covers the whole world. It turns out that since about 1930 there has been a steady increase in the average intensity of the Earth's magnetic field measured at observatories in the northern hemisphere. At the same time the magnetic intensity in the southern hemisphere has decreased.

The changes are more complicated when studied in detail, but correlate remarkably with climatic changes. In Mexico, Canada and the United States, magnetic intensity is decreasing and the climate is getting warmer. Meanwhile, in Greenland, Scotland, Sweden and Egypt the intensity is increasing while the climate is getting colder. This rule seems to apply everywhere around the world.

Sudden changes in magnetic intensity, which are recorded occasionally at some observatories, are followed by abrupt changes in weather. In general Dr Goesta Wollin and colleagues report, the temperature trends lag a year or more behind the changes in magnetic intensity.

So the rule of thumb that a stronger field means a cooler climate works just as well for short periods of time and for local regions of the globe. We still don't know what causes the field to change in the first place. But there is one very intriguing possibility – that changes in the Sun's activity can affect the magnetic field of the Earth. That is a problem right at the frontiers of present-day research, and one that might well make the headlines in the near future. Meanwhile, the evidence relating the overall pattern of the Earth's magnetic field with the general circulation of the atmosphere has been pieced together and publicized in the work of Dr J. W. King, at the U.K. Appleton Laboratory.

Dr King has even coined a new word for this science of the study of weather and magnetism: magneto-meteorology. This is perhaps one of the best examples of the way science has begun to develop over the past few years: it is becoming common for scientists from quite different disciplines to collaborate in tackling major problems that would otherwise have been beyond the compass of any one specialization. The problem of the Earth's climate in general, and magneto-meteorology in particular, has brought together a battery of physical and Earth scientists, and meteorologists, in a particularly constructive way.

Dr King has gathered a mass of data obtained over many years by many different scientists. His starting point is the remarkable similarity between the pattern of the magnetic field of our planet over the northern hemisphere and the pattern of the contour lines of atmospheric pressure. Both patterns have a double 'pole' like a dumbell, and the dumbell shape is lined up in much the same way for both pressure and magnetic distributions. In meteorological terms, the centres of low pressure that dominate northern hemisphere circulation coincide with the centres of greatest magnetic intensity. That is particularly striking because, as Dr King and others have already pointed out (and as we saw in Chapter 5) there is a definite link between the activity of the Sun and the weather in the northern hemisphere. The best explanation of this phenomenon, remember, is that the circulation of the atmosphere is affected by the charged particles from the Sun – the solar cosmic rays – which are focused by the Earth's magnetic field on to a small part of the atmosphere. That suggested link between weather and solar activity becomes even more plausible if key features of the atmospheric circulation are compared with key features of the magnetic field.

It is difficult to see how this apparent relationship between circulation and magnetism could have arisen by chance (see Fig. 13). But the geological evidence tells us the Earth's magnetic field drifts relative to the solid surface of our planet. Is it a coincidence that we are living in an era when the dumb-bell-shaped magnetic pattern just happens to be aligned with the dumb-bell-shaped atmospheric circulation pattern? If this is not a coincidence, then the atmospheric circulation, tied to the magnetospheric pattern, must also have drifted westward during historic times. Is there any

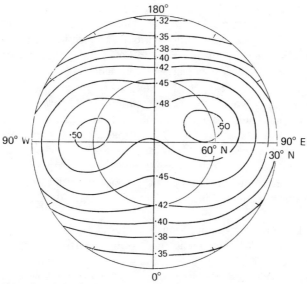

Fig. 13 Top, plot of the distribution of the magnetic field of the Earth for the northern hemisphere in 1965. Bottom, plot of the pressure distribution over the northern hemisphere (presented as contours of average height in decametres of the 500-millibar level in the atmosphere). Are the similarities between the two plots more than a coincidence? (Dr J. W. King, *Nature*)

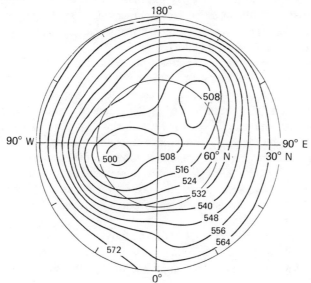

evidence from historical weather information that that is indeed the case?

The answer, of course, is yes. A comparison of Japanese, Chinese, Russian and European historical records mentioned in Chapter 6 already shows hints of some kind of westward drift of climatic change. According to Dr King's analysis, there is equally good evidence even from the records of Europe alone. Movement of the circulation pattern of the northern hemisphere winds would change the nature of the prevailing winds in Europe, and that could make a great difference to the climate. In particular, Dr King points out that the little ice age that affected Britain and Europe some 300 to 400 years ago corresponds to a time when the magnetic declination at London and Paris was roughly zero. If the wind patterns at that time corresponded to the magnetic field of the Earth then, there is every reason for the climate of Europe to have been colder at that time than it is today.

The link with solar activity then becomes clear. Solar cosmic rays are focused by the Earth's magnetic field on to a part of the atmosphere – between latitudes 76°N and 79°N – where a ridge of high pressure is found. This ridge dominates the weather of the northern hemisphere, so that the concentrated effect of the relatively small amount of energy in the solar cosmic rays is enough to cause noticeable changes in the circulation of the entire hemisphere.

One obvious question is: does a similar effect occur in the southern hemisphere? It's difficult to say yet whether this is so, but it seems likely. The circulation pattern of the southern hemisphere is simpler than in the north – there is a single 'pole', not a dumb-bell-shaped pattern – and this is also true of the magnetic field pattern. But of course we just don't have the same kind of extensive historical information from the south as we do from the north. And even today there are far fewer meteorological stations scattered around the ocean wastes of the southern hemisphere than there are dotted around the land masses of the north.

Dr King emphasizes the difficulties associated with developing a reliable prediction system from the science of magneto-meteorology. One of the greatest problems is to allow for the way mountain ranges interrupt the flow of the atmospheric circulation. It may well be that this kind of effect explains why the magnetic and atmospheric patterns of Fig. 3 are not exactly congruent.

Like all new theories, Dr King's ideas have come under attack. This is quite normal, and is the accepted way by which scientific progress is achieved. Anyone who can find apparent flaws in the theory has a duty to point them out; then, if Dr King or someone else can modify the theory to correct the flaws, magneto-meteorology becomes stronger and more reliable. Of course, if someone were to find flaws that could not be rectified then the theory might have to be abandoned.

Towards the end of 1974, for example, Dr J. S. Sawyer, of the U.K. Meteorological Office, made some specific criticisms of the concept of magneto-meteorology, and King responded to those criticisms in what seems to be a satisfactory manner. It is worth looking at the details of that dialogue, as an example of how science progresses, and as a reminder that progress is indeed being made, even when we have the sometimes confusing spectacle of specialists disagreeing with one another.

There is no real doubt that long-term changes in the circulation of the lower atmosphere are accompanied by climatic changes. Where Sawyer took issue with King was on the question of whether the geomagnetic field causes these changes. He said that 'it is unnecessary, and probably misleading, to postulate any causal relationship between the geomagnetic field and the [pressure] contours, because the main features of the ... contours can be explained by direct calculation from physical and dynamical principles without consideration of magnetic effects.'

Now, the Met. Office itself has pioneered, and continues to work on, the very important field of numerical modelling – that is, simulating the movement of the atmosphere over the solid Earth using sophisticated mathematical models and the latest high-speed electronic computers. Stripped of its scientific jargon, Sawyer's statement really boils down to saying, 'Our idea can explain everything, so we don't need your theory.' However, King has not said that magneto-meteorology can explain everything – he even pointed out how essential it is to use the numerical modelling to explain why the Earth's atmospheric circulation is not perfectly congruent with the magnetic pattern. Before Sawyer made these specific criticisms, King had already said that the magneto-meteorological evidence 'cannot be construed as proof that the Earth's magnetic field and the average tropospheric pressure patterns are related, but it does suggest that more work on this problem should be undertaken.'

A lot of meteorologists certainly would not agree that present computer models can explain everything without invoking a magnetic field effect. Indeed, the different models developed at different research centres (numbering about a score in all at the end of 1974) differ from each other, and each one differs somewhat from the behaviour of the real atmosphere. Clearly, there are effects which are not allowed for in these models. And one of them could well be the influence of magneto-meteorology. But according to King some of the models may already include the effects of magnetic fields, at least partially and in a roundabout way. The temperature maps which are used as a basis for the mathematical models are, of course, based on observations of the real temperature distribution over the surface of the globe. And if magneto-meteorology is important to the global climate, these basic temperature maps are already distorted by the effects of the Earth's magnetic field!

In order to confirm the theory, the question 'Are the magnetic field and the weather related?' must be answered conclusively. The best hope of this seems to be from comparing detailed studies of meteorological data obtained from records going back over many years with reliable maps of the magnetic field. The accurate magnetic information can almost certainly only come from geological studies of the fossil magnetism of rocks; the corresponding weather maps must come from historical climatology of the kind mentioned in Chapter 6, where meteorological information is inferred from agricultural and other historical records.

This would require scientific co-operation on a grand – and desirable – scale. But at present the different pieces of the climatological jigsaw puzzle are still only beginning to be fitted together to form a recognizable picture. Some of the pieces may yet prove flawed, and will fail to stand up to the rigorous and all-important criticism of other scientists. Others will turn out to be key pieces of the puzzle, from which great new advances can be made. And, in between, the vast majority of the pieces will be more mundane additions to the store of scientific knowledge.

8 Ice and Fire

The present interest in climatic change might make it seem that this is a completely new branch of science, but that is not really the case. Of course, practical meteorology has only been possible for a hundred years or so, as we have seen. And for a good fraction of that hundred years it was generally believed that the climate did not change – so there was very little study of climatic change! But ideas such as magneto-meteorology and studies of the influence of solar cycles on the weather perhaps give a false impression of the true strength of this branch of the meteorological sciences. These new sensational – and to some extent still controversial – ideas are not the sum total of climatic research, and indeed many climatologists would say that the publicity they have received gives a false impression of what is really a solidly based science which has, for decades, been making steady if unspectacular progress.

One place where this progress has been recorded in a very complete form is at the Voeikov Geophysical Observatory in Leningrad. In a manner rather reminiscent of the Soviet space programme, the climatologists in the U.S.S.R. have tried to develop solid, reliable themes in meteorology over a period of decades, building in a steady, logical way upon the work of previous generations. They are not unique in this – many other climatologists around the world are carrying out their research in much the same way. But the study of climatic analysis based on the laws of conservation of mass and energy as applied to the atmosphere and the surface of the Earth has probably been carried out more comprehensively at the Voeikov Observatory than at any other single laboratory in the world, and this study has recently been reported in great detail in a book written by the Director of the Observatory, M. I. Budyko.* Of all the studies of

Climate and Life (Academic Press, London, 1974).

climatic change, the work at the Voeikov Observatory probably provides the best single insight into the 'traditional' approach.

In this approach, the two key influences on the energy–balance equations are polar ice and the dust thrown out into the atmosphere from the fiery mouths of volcanoes. The method is straightforward in principle; to calculate the balance between the heat received by the Earth from the Sun and the heat radiated away into space. But in practice that is far from easy.

The actual radiation at the Earth's surface depends on the temperature, the amount of water vapour in the air and the cloudiness; it can vary from zero to several tenths of a calorie per square centimetre per minute. Oceans and other bodies of water play an important part in the energy balance, because of the different effects of liquid water, ice and water vapour in the radiation, or absorption or reflection of heat. The beginnings of the energy–balance approach to climatology came in the late nineteenth century, when Voeikov (after whom the Leningrad Observatory is named) and others pioneered this kind of work. From the 1940s this work has made great strides, and today maps of the annual variations of the heat-balance of most parts of the world are available. Climatologists are well on the way to understanding the present heat-balance of the globe – and that, of course, is the essential first step towards explaining past variations in climate.

It has taken so long for the study to reach this stage (the best part of 100 years) simply because the Earth is so big, and until recently communications on a global scale were so difficult. At present, there are about 1,000 stations around the world making radiation and heat-balance observations. Even today this network is far from satisfactory. There are very few stations in the developing countries, and observations from the great areas of the Earth covered by sea are limited to weather ships. As with so much climatological work, there is an urgent need for more observations, and that means more funds.

But there is one part of the world which is fairly comprehensively studied by the Soviet climatologists, and that is the Arctic. That area is particularly important for the worldwide circulation and weather patterns, and of course the U.S.S.R. is just about ideally situated to study it. With the new tool of satellite photography (see Chapter 4), combined with the more traditional approach of the Soviet teams, the significance of the polar ice for the climate of the whole world is becoming clearer. In the geological past, ice cover has varied a great deal. During the Quaternary period (the past million years) the ice cap has repeatedly spread to middle latitudes over the sea, with corresponding glaciation on the continents; and during the warmest periods between glaciations ('interglacials') the ice cap may have disappeared altogether. Even during the past thousand years the movement of the ice has had a profound effect on human history, as we have already seen. The climatologists at the Voeikov Observatory have been paying particular attention to the twin problems of how climatic changes affect the ice cover, and how the changes in ice cover can in turn affect the climate.

Ice begins to form on the surface of the sea when it cools to a temperature of $-1.8°C$. Once ice has formed, it will grow thicker as long as the temperature of the surface is lower than the temperature of the water at the bottom of the ice; the rate of freezing depends on how quickly heat can be transferred through the ice. That depends not only on the temperature difference but also on the thickness of the ice, and whether or not it is covered by snow. All this makes the study of the influence of ice cover on climate far from simple. But considerable progress has been made in these studies, with fascinating results.

The simplest example of the effect of snow and ice cover on climate is the hypothetical situation of an Earth completely covered by snow and ice, and with an atmosphere completely free from clouds. That case is simple enough for exact calculations to be carried out easily, and it turns out that, allowing for the greater reflectivity of ice and snow compared with the Earth as it is today, there would be a reduction of the average temperature by 75°C. That applies to the whole atmosphere, and in fact at the Earth's surface the temperature in such a situation would be as low as $-87°C$, about 100° less than the average surface temperature today, which is roughly 15°C.

The meteorologists at the Voeikov Observatory have calculated just how much effect the ice cover over the Arctic Ocean alone has on the global climate in general, and on the temperature in the Arctic in particular. It seems that if all other climatic factors remained the same, and in particular as long as the heat input from the Sun were constant, the effect of removing all the Arctic ice today would be to raise the average temperature of the Arctic regions by more than 40°C. Now that is immediately intriguing, because such a rise in temperature would be enough to stop the ice re-forming. It looks very much as if the heat-balance of the Earth could quite happily adjust to a warmer situation, in which the Arctic Ocean would be completely free from ice. And that would have important repercussions on worldwide weather patterns.

The first detailed calculations of this effect were made only in the 1960s. They confirm that the Earth can exist in either one of two climatic regimes corresponding to the presence or absence of ice at high latitudes. In addition, the calculations show that *both* of these situations are basically unstable; because of this, slight warming or cooling produced by outside influences would be enough to reverse the situation, with ice forming or disappearing very rapidly as a result of the slightest changes in the factors which determine the global climate. This certainly agrees in a most disturbing way, with the satellite observations of how ice cover has changed in recent years (see Chapter 4).

A lot of work on this problem is now going on outside the U.S.S.R., of course, and in 1966 the first international meeting to look into this problem took place in Los Angeles. One suggestion made at that meeting was that the climate at high latitudes during ice-free periods would be typified by cool summers and mild winters, with heavy precipitation. Curiously enough, it seems this could, in some circumstances, lead to extensive glaciation in places like Europe and North America, even while the Arctic Ocean itself remained more or less free from ice. But this possibility has not really been followed up. The Soviet studies concentrate on the more logical situation, when a rise in temperature produces a general melting of the glaciers.

The sensitivity of the balance between the two kinds of ice-cover situations is highlighted by calculations of the temperature of the sea and air in the Arctic under ice-free conditions. According to the best estimates, the sea temperature would range between 4.3°C (summer) and − 0.8°C (winter), with corresponding air temperatures of 5.8°C and − 5.4°C. In other words, the Arctic Ocean in winter would be only 1° warmer than the freezing point of salt water. But over a whole year, the average temperature in the central Arctic under ice-free conditions would be about 15°C higher than the average temperature today. Such a temperature rise would not be uniform throughout the year. Temperatures in winter would be much higher, but temperatures in the warm season would be only slightly higher than those experienced at present. There seems little doubt that a similar rise in temperature, although not quite so pronounced, would spread from the edges of the Arctic regions to middle latitudes, and perhaps even into the tropics.

This change in temperature patterns would produce a more uniform situation, with weaker temperature gradients and, as a result, weaker circulation patterns in the atmosphere. One effect of this kind of change would be to cause a redistribution in the rainfall patterns of the world. One study at the Voeikov Observatory, made in the mid-1960s, suggests that coastal regions might get more rain, but that in many interior regions the amount of precipitation would decrease.

There would also, of course, be a marked rise in sea level because of the melting of glaciers on land. That could cause extensive flooding – and two of the cities most affected would be London and New York. So, although the idea of a warmer Earth might be attractive in some ways, such a situation would cause considerable problems. Is such a change likely? What could cause it, and how might we prevent it occurring? These questions have also been studied by the Soviet climatologists, who have focused attention on one or two of the important factors in reversing the climate.

Just how much do climatic conditions on Earth today differ from those that have been 'normal' for the past few hundred million years? Throughout most of that time, the difference in temperature between high and low latitudes was much less than it is today. Low latitude temperatures have always been much as they are today, but the geological evidence shows that temperatures in middle and high latitudes were generally rather higher than they are today: similar to our estimates of what the Earth would be like if there were no ice in the Arctic Ocean. In itself, this does not prove anything, but already the evidence is suggestive.

Though present conditions are unusual in the long history of our planet, they have been pretty persistent by a human time-scale. The evolution of the present pattern of contrast between the equator and poles seems to have begun about 70 million years ago, at the beginning of the Tertiary period. But the situation only changed slowly, and by the beginning of the Quaternary (roughly a million years ago) conditions were still very different from today's. But then there was a sharp drop in temperature at high latitudes, and geological evidence shows the development of extensive glaciation during the Quaternary. This glaciation has fluctuated considerably, sometimes reaching well into middle latitudes, but since the so-called Würm stage, which ended about 10,000 years ago, it has been largely restricted to the Arctic Ocean, mountain regions, and islands at high latitudes. Even during the past 10,000 years, however, the ice cover has fluctuated. These fluctuations have been on a smaller scale, but enough to affect man's activities and the development of civilization.

As we have already seen, during this century we have had a warming trend, which reached a peak in the 1940s and has since been followed by a cooling trend, which has not yet caused enough cooling to counteract completely all the effects of the warming trend. The only certain thing about all these changes is that as yet there is no generally accepted explanation of the causes of climatic change and fluctuations. But there are some pretty good pointers.

It certainly is possible that something odd happened to the Sun about 70 million years ago, which caused the very long-term change in climate since then. This would tie in with the speculations made by Professor Fowler and others, which we discussed in Chapter 6. But the practical use of such theories is slight. Shorter term effects, connected with sunspots, and so on, are clearly of greater importance and of more immediate practical use. It's quite helpful to know what is likely to happen over the next 11 years, but not much help to know what might happen in a million years' time.

The fundamental assumption behind the work at the Voeikov Observatory is that the heat arriving at the top of the atmosphere from the Sun is constant. That is a reasonable working assumption, but it means that the results which come out of this work must be combined with studies of solar fluctuations, and the influence of sunspots and so on, to produce a more complete picture of the causes of climatic change. But that further step has yet to be carried out in detail.

If we assume that the Sun's influence is steady, what could affect the amount of heat arriving at the Earth's surface, and thus disturb the heat balance? The answer is pretty obvious – changes in the transparency of the atmosphere, either through changes in cloudiness, or through changes in the amount of atmospheric dust. Dust changes have received a great deal of attention from the Soviet climatologists – and also from other experts, notably Professor Hubert Lamb in England – and these studies have shown just how important volcanic activity might have been, over the millennia, in determining the heat balance of the world. The amount of radiation reaching the surface of the Earth does vary from year to year, even allowing for the effect of cloud cover. There are also longer changes, over periods of decades, which have been recorded by observing stations since studies of this kind began at the end of the nineteenth century. Many of these changes can be directly related to the occurrence of explosive volcanic eruptions, which flood the lower part of the atmosphere with fine particles of dust for a short period of time.

After large volcanic outbursts, the amount of solar radiation reaching the surface of the Earth can fall over a couple of months by a remarkable 10 or 20 per cent, and the subsequent clearing of the atmosphere can take as long as a year, and in some cases even longer. Most people have heard of the famous volcanic outburst at the end of the nineteenth century, when the island of Krakatoa exploded; the explosion had a dramatic effect on the transparency of the atmosphere to the heat radiation of the Sun. It happened at just about the time when reliable monitoring of the heat received from the Sun was getting established. Since the end of the 1880s, a graph of the average amount of heat received under cloudless skies in the northern hemisphere shows two peaks – a sharp maximum just before 1900, then a slight dip rising to a broader maximum running from 1920 to the mid-1940s. We are now experiencing another dip in this curve – in other words, relatively little heat from the Sun is penetrating the atmosphere. It might well be that the increase in radiation at the end of the nineteenth century corresponded to the clearing of the atmosphere after the spread of volcanic dust from the explosion of Krakatoa.

Applying this idea to the changes in measured radiation during the twentieth century, the Soviet climatologist M. I. Budyko has suggested that the reduction in radiation just after the turn of the century could have been the result of eruptions of Mount Pelée and other volcanoes, and that the rise in measured radiation just before 1920 corresponds to a clearing of the atmosphere after the explosion of the Katmai volcano. For a long time after that, he says, no great explosive eruptions occurred.

What of the decline since the 1940s? Some people have suggested that this might be caused by the pollution of the atmosphere caused by man's activities. But even today man's industrial pollution cannot produce such dramatic effects on the global atmosphere as natural phenomena do – and eruptions of Mount Spurr in Alaska (in 1953) and of Mount Bezymiannyi in Kamchatka (in 1956), together with other volcanic activity, may have played an important part in establishing this new trend.

There is no doubt that changes in atmospheric transparency do affect global temperature. A change in radiation absorbed by the Earth of just 1 per cent can, according to the latest calculations, produce a change in surface temperature of between 1.2°C and 1.5°C. There is another step in the calculation, because the dust scatters some radiation sideways, where it can still be absorbed by the atmosphere. Not all of the heat is lost back into space. But, according to Budyko, a volcanic eruption which causes a 10 per cent reduction in the amount of direct radiation measured at the Earth's surface corresponds to a loss of global radiation of about 1.5 per cent – enough to cause a reduction in temperature of 2°C. Even this is not quite the end of the calculation, since a further correction is needed to take account of the different effects of land and water, and of the movement of the atmosphere. All this reduces the effect of the dust still further, and Budyko ends up with the conclusion that an eruption causing a 10 per cent drop in direct measured radiation will cause a fall in global temperature of just a few tenths of a degree. That corresponds very well with the variations in temperature which have been measured over the same period (roughly 100 years) for which direct radiation measurements are available. All in all, it seems quite plausible that large volcanic eruptions make a contribution to the variation of the average annual temperature of the globe.

The Soviet view is that, although other factors cannot be excluded, 'they evidently play a less significant role, when compared with changes in atmospheric transparency.' I feel this view is limited, since it assumes that the Sun is providing an unvarying source of heat. The influence of the Sun over the 11-year cycle and longer periods is now unquestionable, and the new science of magneto-meteorology provides one attractively plausible indication of how this influence might work. In fairness, though, this work is so new that we do not yet know the Soviet climatologists' reactions. I believe that these two effects – the solar influence and the changes in atmospheric transparency – are probably the most important factors affecting climate and contemporary climatic change.

This does not mean that all the work on other aspects of the problem is not important. We must understand how the circulation of the atmosphere and the ice cover of the Arctic, for example, are related to the changing climate. Indeed, in terms of direct relevance to the climate of next year, or the next four or five years, those studies are just as important as the fundamental questions of what causes the climate to change. All the evidence has to be fitted together before we can see how volcanic explosions in the Far East, or slight disturbances on the Sun, lead to poor harvests in the great plains and cause disruption of oil prospecting in the North Sea.

Studies of climates of the geological past also help to provide an understanding of the workings of the weather machine, but they involve such long-term changes that they do not offer much help to our immediate problems. A few highlights are, however, worthy of mention. A relatively small change in radiation – no more than 1 or 1.5 per cent – would be enough to spread ice cover over both land and sea down to the middle latitudes. Such a change might, of course, be the result of a slight hiccup in the Sun's output. But it is also interesting that the changes necessary are only a few times greater than those that have been measured in this century as a direct result of volcanic activity.

So volcanic activity alone might well play a significant part in the spread of glaciation. This idea gains strength from geological studies which indicate that the main epochs of glaciation during the Quaternary (the past million years or so) seem to correspond to times when there was substantial volcanic activity. The obvious question is why similar levels of volcanic activity in the long period of geological time preceding the Quarternary did not cause similar glaciation. One answer could be that the unusual conditions that are found around the North Pole today simply did not exist until some tens of millions of years ago.

As we have seen, the importance of present-day conditions is that the Arctic Ocean can exist quite happily either as an ice-covered or as an ice-free sea. Because of the drift of the continents, and changes in the sea level over geological time, the Arctic Ocean has not been landlocked for most of the Earth's history. With better circulation of warmer water from lower latitudes into the region of the North Pole, the 'Arctic Ocean', or its counterpart, could never freeze. Of course, when the North Pole was covered by land there must have been periods of more intense glaciation – just as the presence of Antarctica surrounding the South Pole today has a profound influence on the climate of the southern hemisphere. But that really is another story.

Before leaving the topic of volcanic dust altogether I cannot resist mentioning one speculative possibility. As I have discussed in another book,* it is now pretty clear that variations in the interplay of forces between the Sun and Earth over the solar cycle can affect such phenomena as earthquakes and volcanic activity. There is very little doubt that effects associated with sunspots could also trigger volcanic eruptions. This is no more than a speculation at present, but one that perhaps needs further investigation.

---

* John Gribbin and Stephen Plagemann, *The Jupiter Effect* (Macmillan, London, 1974; Walker, New York, 1974).

9 Are Man's Activities Changing the Climate?

The quick answer to this question is, of course, 'yes.' By cutting down forests, burning fossil fuels which release heat and smoke, and flying aircraft in the stratosphere, mankind must be influencing the workings of the atmosphere to some extent. But how much?

Just as meteorologists fifty years ago regarded climate as 'average weather', and thought of it as essentially unchanging, so most people until very recently have thought of the atmosphere of our planet as some kind of infinite reservoir, capable of absorbing the worst of our polluting activities without harm. Now, it is fashionable to hold the opposite view, with the atmosphere regarded as a frail shell which should be protected in every way possible. But both of these are essentially emotional points of view. The first thing we need to know before we can take a cool appraisal of just how much damage man's activities might be doing to the climate is just how delicate the balance of climate is. So studies of past climates, and how rapidly they have changed through natural processes, can provide a good background.

Some of the best recent studies of sudden changes in climate over the past 10,000 years (the Holocene period) have been carried out by Professor Reid Bryson and his colleagues at the University of Wisconsin–Madison. Bryson's group has pioneered the use of pollen samples from sediments and lake varves as indicators of how the climate has changed – and its discoveries have surprised a lot of people.

It has become clear in recent years that climate changes in abrupt 'steps' – short periods of transition which separate long periods of relatively stable climate. One way of establishing the occurrence of these steps is by recording variations in such things as ice cover, and the extent of certain kinds of vegetation. These indicators are sometimes termed 'climatic proxies'; an abrupt change in the 'proxies' at some date in the past provides good evidence that the climate changed suddenly at about the same time. But this technique has only really been practicable since the carbon-14 dating technique was established. Radiocarbon is absorbed by all living things up to the moment they die. Then, the radioactive carbon-14 decays, at a known rate, into non-radioactive carbon-12. By measuring the proportion of radiocarbon left in pollen or other organic debris found in sediments, it is possible to calculate fairly accurately how long it is since the sediments were laid down. And the climatic discontinuities are thus recorded.

Bryson argues that sudden 'steps' in climate must have a significant impact on human civilizations. In some of the pioneering work at the University of Wisconsin–Madison, carbon-14 dates of variations in pollen from such sediments were measured. Then, the climatological team gathered geological evidence providing estimates of the dates of the beginning and end of glacial periods (related from changes in the height of sea level around the world) and the dates at which peat beds were laid down (again calculated from radiocarbon measurements). Finally, in a complementary study, the dates at which dramatic changes in human cultures took place were compared with these dates, to test the assumption that abrupt changes in climate must have a profound effect on human activity and civilization.

Because the object of this kind of investigation is only to find *sudden* changes, the problem is simpler than it might seem at first sight. In the botanical analysis, the dates used were those corresponding to the greatest or least deposits of each pollen species, and the dates used from the peat deposit record were only those corresponding to the beginning and end of each phase of peat deposit. In the same way, only the extremes of glaciation were noted, and the greatest and least heights of sea level from the geological record of the past 10,000 years. This gave Bryson's team a total of 815 measurements from various sources. All of these sources were scientific papers published by specialists working in the various areas of study, often on problems completely unconnected with the study of climatic change. Thirteen 'significant' dates were revealed in this way, and all but one of them could be picked out in the pollen data alone, So for the next, and more sophisticated, stage of the analysis Bryson's group used just the pollen data, a total of 630 radiocarbon measurements which contained the 12 significant dates.

Then the pollen data were examined using strict statistical tests, and six dates picked out as being of greatest importance: 850, 1680, 2760, 5060,

8490 and 10,030 years Before the Present (B.P.) The pollen data came from sites covering a large part of the northern hemisphere, and so should provide a good guide to genuine changes in the global climate, not just to local weather effects. So how do they relate to the cultural record of man's activities? It seems likely that cultural patterns will end, or change noticeably, at times of sudden climatic change. Once again, radiocarbon dating of the remains left by man provides a straightforwward guide to *sudden* changes in his pattern of behaviour, although it would be far harder to use them to study *gradual* changes. The most significant dates that emerged from the Wisconsin team's study of variations in the pattern of man's activity across the northern hemisphere were 830, 1260, 2510, 3110, 4230, 5900 and 9530 years B.P.

The agreement with the botanic data is not exact, but it is certainly interesting. The botanic discontinuity 2760 years ago and the cultural change 2510 years ago are each the most important in the two kinds of data, and improved climate could well have played a part in the development of Bronze Age culture in Britain, for example. The Bronze Age lasted from about 2000 B.C. to 500 B.C., and a minor climatic deterioration which occurred roughly in the middle of this period may have played a significant part in establishing the move towards hill forts and settlements, according to archaeologist Colin Burgess.* Coming closer to the present, the botanic/climatic change towards a harsher regime some 850 years ago coincided with many cultural changes across the northern hemisphere, such as the decline of the Vikings as a seafaring nation and changes in the culture of the inhabitants of the Mill Creek region of northwestern Iowa. With dates of less extreme changes in cultural and botanic data included, Bryson and his colleagues find more agreement between the two sets of dates. But this is still a new field of study, and not too much can be made of the relationship at present. Even so, some cultural changes, at least, do seem to have been triggered by sudden climatic change. In present-day terms, however, the most remarkable (and worrying) thing to emerge from this kind of analysis is just how quickly important climatic changes can take place.

Bryson says that the pollen records indicate 'sharp' changes in climate occurring over just 17 years on occasions in the past; when pressed to

*See Chapter 5 of *British Prehistory,* edited by Colin Renfrew (Duckworth, London, 1974).

elaborate on what he means by 'sharp' he goes on to say that it is possible, judging from the pollen record, for the climate of our planet to change from an interglacial situation – like the present climate – to a full glacial situation in just 100 years. That is certainly a dramatic and even frightening possibility. Many meteorologists say that, according to their theories of the physics of climatic change, such a sudden decline in climate is impossible. But Bryson says that the pollen evidence is incontrovertible, and that 'the theories must be made to fit the facts, not the other way around.'

If such sudden changes have occurred in the past, then they can occur

Fig. 14 An example of man's influence on the atmosphere. This cirrus cloud grew up over Colorado from the vapour train (contrail) of a high-flying jet aircraft. (N.C.A.R.)

again. And if the atmosphere is so delicately balanced, then it begins to look very much as if we ought to be worrying about the possible influence of man's activities.

Bryson is now doing just that. Starting out from the idea of volcanic dust as the root cause of ice ages, as described in Chapter 8, he has studied man's polluting activities in terms of the amount of dust introduced into the atmosphere. Since the middle of this century there seems to have been a more or less steady fall in temperature, and Bryson relates this to a more or less steady increase in the amount of dust in the atmosphere, which he

Fig. 15 Near Denver, Colorado, steam released by a power-generating plant creates a small cumulus cloud. (N.C.A.R.)

blames on the growth of man's industrial activities. Over 20 years, he says, we have gone 1/6th of the way towards a new ice age – a rate of change which ties in uncomfortably closely with his estimate that 100 years can see a transition from interglacial to full glacial.

But – and it is a big 'but' – Bryson's views are far from being shared by all his climatological colleagues. His figures have been attacked on several fronts. Firstly, it is not absolutely certain that the Earth really has been cooling since 1950, because some of the standard meteorological sites where temperature is measured have been changed in the past 20 years. Another problem is that those sites might just possibly be recording local cooling changes which are taking place in spite of a warming up of other parts of the globe, where there are less recording stations. But most meteorologists agree that there is a real decline in temperature, although they disagree on how great a decline it is. Secondly, there is some dispute about whether or not there is an increasing amount of dust, man-made or otherwise, in the atmosphere. Dr Stephen Schneider, of the National Center for Atmospheric Research (N.C.A.R.) in Boulder, Colorado, pointed out to me that Bryson's estimates of how dust in the atmosphere has increased depend on measurements made at only one or two sites. With so few sites, the observations might very well be unusual. They can't provide a global view of what is happening to the transmissivity of the atmosphere. Dr J. S. Sawyer points out that one of the sites used for the measurements on which Bryson's calculations depend was affected by dust from a nearby volcano, and that this was only realized after Bryson's study was publicized.

Finally, there is still some doubt about whether man-made dust will cause the atmosphere to reflect more heat from the Sun or to absorb more heat. In simple terms, 'grey' dust over a white snowfield would actually cause an increase in heat absorbed, while the same 'grey' dust over dense forest might cause a decrease in heat absorbed. It all depends on the nature of the dust and of the underlying surface, problems which have not been tackled on a large scale.

Nothing daunted, Bryson points to other worrying aspects of man's interference with his environment. The clouds which built up from the vapour trails of high-flying jet aircraft, for example. Over the North Atlantic, the world's busiest jet route, these clouds can cause a significant loss of sunlight and heat, says Bryson. The result is that the sea below is less hospitable for the micro-organisms, called plankton, which are at the bottom of the food chain of the oceans. (Some fish eat plankton, other fish eat those that eat plankton, and so on, upto and including man, who eats many kinds of fish.) And Bryson says that a recent study by the Scottish Marine Biology Association shows that there has been a decline in the amount of plankton and of herring in that part of the Atlantic which lies just under the jet routes.

Dr Sawyer deplores the current situation where it is becoming almost a competition to see who can suggest the latest possible influence of man on the climate. With the limited resources available to meteorologists, and the limited number of climatologists available, this is rather futile since it is impossible to investigate properly all of the suggestions raised.

Nevertheless, there are three topics which deserve immediate study, and which are receiving it. These concern the possible effect of all the carbon dioxide we are producing by burning fossil fuels (coal and oil); the effect on the climate of tiny dust particles which remain in the air for a long time (aerosols); and the possible direct effect of all the heat produced by man's activities. There is also a related problem, which is not strictly climatological: the effect of the gases used in spray cans on the ozone layer of the stratosphere.

At one time, the carbon dioxide problem caused a lot of concern. If there is more carbon dioxide in the atmosphere, more of the heat being radiated by the sclid Earth and the oceans is absorbed in the atmosphere and then re-radiated. Since the re-radiated heat goes out in all directions from the middle of the atmosphere, some of it gets back to the Earth's surface and warms it up, in a process known as the 'greenhouse effect'. This effect is undoubtedly a real one – but how much warming is it likely to cause? According to the latest calculations, we have nothing to worry about from the greenhouse effect. If we carry on using more and more fossil fuel each year, at an exponential rate of growth, then by the year 2000 the effect of all the extra carbon dioxide will be to increase the average temperature of the globe only by a small fraction of a degree Centigrade. As we have been told only too often recently, unlimited exponential growth of that kind would mean that all of our reserves of fossil fuel would be used up by the end of the century – so after that tiny rise in temperature there would be nothing else to worry about! Of course, more sensible use of fuel would stretch our available resources well into the twenty-first century; but in that case there will be less carbon dioxide put into the atmosphere each year, and the greenhouse effect will do even less harm.

The effect of aerosols is not yet so well understood. Big particles of dust are no great problem in climatic terms; they are more easily removed from industrial smoke than small particles, and in any case those that do escape are soon washed out of the atmosphere by rainfall. Dr John Firor, the Executive Director of the National Center for Atmospheric Research told me that a recent repeat of an experiment originally carried out in 1910 showed that there has been no increase in the overall dust level in the atmosphere in the past 60 years; but a new experiment comparing the atmosphere in the countryside with the air in towns showed that while cities have become cleaner over the past 10 years the countryside is becoming dirtier. So there is more to the story than suggested by a simple average, and what the details tell us is that big particles are indeed being removed but that many small particles remain.

As well as the problem of calculating reflectivity – are we talking about grey clouds above a black surface or grey clouds above a white surface? – there are many other unanswered questions about the influence of aerosols on climate. These tiny particles may, for example, act as 'seeds' for the growth of water droplets, and encourage the spread of new clouds. That would increase the Earth's overall reflectivity and make it cooler. On the other hand, the dirt particles may just stick to water drops and make existing clouds dirty. In that case, the clouds will be darker and will absorb more of the Sun's radiation, causing the Earth to warm up. Stephen Schneider emphasizes that this kind of study is really only just beginning and that the answers to these puzzles are not 'just around the corner'. There could be a problem of man's influence on the climate here, but we can't say yet just what that influence is: which suggests that Bryson's warning of dramatic change may perhaps be exaggerated.

Studying the effect of the heat resulting from man's use of energy provides some more clearcut answers, at least in the sense of indicating general tendencies. The effects are not too different from those produced by changing forests into fields, which have different reflecting properties, and most climatologists do not seem greatly concerned by them. As of 1970, the amount of heat produced by man corresponded to about 1/15,000 of the energy absorbed from the Sun. The time to begin really worrying about the influence of man's heat input has been put by some climatologists as the time when this input reaches 1 per cent of the solar input, and that is not due for 130 years at a growth rate of 4 per cent starting from the 1970 figures. It's worth noting that this heat, of course, is independent of the fuel supply; even the as yet hypothetical nuclear fusion reactors produce heat (that is the whole point of trying to build them!) so we can extrapolate far into the future, after the time when coal and oil have been exhausted and nuclear or other power has come to dominate man's energy supply. At the very least, the calculations show that the relevant time-scale is so long that the climatologists have time to study this aspect of the problem carefully and work out what results we can expect from this influence of man on his environment.

Some of the studies of the greenhouse effect which have been made using computer models highlight the present difficulties of this kind of work. One estimate suggested that doubling the amount of carbon dioxide present in the atmosphere would cause a rise in temperature of 10°C. But this rise in temperature would cause increased evaporation of water from the oceans, and produce more cloud cover. And a 1 per cent increase in cloudiness would cause enough extra reflection to cancel out completely the carbon dioxide greenhouse effect! Other calculations differ from that example in detail – one suggests that a doubling of carbon dioxide would cause a 2.4°C rise in temperature and that this would be 'masked' by a 3 per cent increase in cloud.

With that caution in mind, the effect of propellant gases on the ozone layer which protects us from the Sun does look to be a very worrying one at present, even if it is not strictly a climatological problem. These gases are used in just about all spray cans – they are the gases under pressure which push out the hairspray, deodorant, fly-killer, whipped cream or whatever the 'useful' contents of the can may be. These gases are being released into the atmosphere in ever-increasing numbers.

Ozone is a form of oxygen in which the molecules contain three atoms instead of the usual two. It builds up where sunlight has split up ordinary molecules of oxygen, providing spare single atoms which join on to other molecules (mainly high in the Earth's atmosphere). Calculations by two scientists at Harvard University show that although these propellant gases are inert chemically in the laboratory, they can act efficiently to break down ozone in the upper region of the atmosphere.

The importance of all this is that the ozone layer of the stratosphere acts as a shield for the Earth's surface, protecting us from the strongest ultra-violet radiation from the Sun. Even if that shield is not competely destroyed, a reduction in ozone could let enough ultra-violet radiation through to cause widespread occurrence of skin cancer and other effects. According to the Harvard calculations, a 10 per cent growth in the release of propellant gases each year would lead to a reduction in the ozone layer by 15 per cent by the end of the century – and at present the use of these gases is growing at a rate of more than 20 per cent a year. This kind of activity of man – like the use of supersonic transport aircraft, which might also harm the ozone layer – is clearly to be avoided. We have to have power stations, so some increase in heat input to the atmosphere and in carbondioxide levels is inevitable. Fortunately, those effects don't seem too hazardous. But we can all manage without spray cans and without flying through the stratosphere faster than the speed of sound, so why take chances with these dangerous possibilities?

Getting back to the strictly climatological effects of man's impact on the environment, the immediate future does not look quite as bleak as Reid Bryson suggests. The balanced view of most meteorologists certainly does not give cause for complacency, but it does suggest that we have at least got time to investigate the possible hazards and find solutions.

There is some uncertainty about how man's activities will affect the

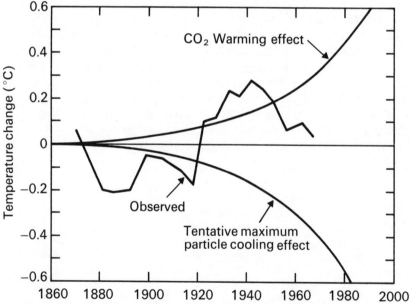

Fig. 16 Just how are man's activities affecting climate? At present it is difficult to say which of two opposite effects dominates, as this plot produced by U.S. climatologist Murray Mitchell shows. The warming 'greenhouse' effect due to increased carbon dioxide and cooling effect due to increased dust seem to just about cancel out – and actual temperature changes during the past 100 years are too large to be explained by any man-made influence alone.

climate over the 50 years or so. Will the increase in carbon dioxide and the greenhouse effect cause a slight increase in temperature (other things being equal), or will the change in the reflectivity of the atmosphere caused by the presence of aerosols lead to a significant reduction in global temperatures? According to some estimates, because of the different time-scales over which these two effects operate, we might well experience first a warming up and then a slightly more pronounced cooling down.

In the longer term − 100 years or more − man's activities are likely to have a more serious effect through the heating up produced as a result of our use of energy. Looking at what has been achieved over the past 100 years, it is difficult not to believe that if our technological civilization survives for another 100 years we will not have at least some kind of an answer to this problem. And if our technological civilization collapses, then of course this particular problem collapses with it.

Stories of imminent climatic doom make good newspaper headlines, and the Reid Brysons of climatic research will no doubt continue to get their share of those headlines. But among his peers Bryson is respected not for his predictions about future climate but for his important discoveries about past climates and how they have changed. In this context, man's activities seem among the least of our problems, for remember that, as Bryson has shown, the climate can change quickly and dramatically without any intervention from man. The pressing question seems, therefore, to be just what kind of natural climatic change we are now experiencing, how extreme it is likely to be and how long it will last. There is no real prospect of deliberate modification of the global system to improve the climate: we must find out how the climate is changing naturally and learn to live with those changes.

10 How Real Is the Threat of Ice?

During 1974, while this book was in preparation, scientific opinion about the development of ice ages was revolutionized. Several developments in different but related areas of research came together, as a result of work during the 1960s and early 1970s, to show that ice cover over the northern hemisphere can develop much faster than was thought previously; in addition, it has now become generally accepted that a situation more icy than the present is more normal for the Earth, judged on the evidence of the past few hundred thousand years, and that as many as 20 full ice ages have occurred in the past 2 million years.

For once, the public has been kept fully in touch with new scientific developments almost as they were happening. A TV special and associated book called *The Weather Machine** broke the news of what the author, Nigel Calder, has called 'the threat of ice' to a mass audience in a dramatic fashion.

Certainly the threat of ice should not be dismissed entirely, and must clearly be an extremely grave problem within a time very short compared with the history of our western civilization. But it is not the most serious aspect of the changing climatic pattern that we have to worry about over the next ten or even twenty years. A close look at the causes of this threat shows just why the worst fears of the 'climatic doomsdayers' may be a little exaggerated.

I have already mentioned the way in which studies of pollen samples in sediments from lakes and oceans (and on land) now reveal how quickly the Earth can switch from an interglacial situation to a full ice age. Studies of cores drilled from ice sheets such as those which cover most of Greenland

*B.B.C. Publications, London, 1974.

can also give an indication of past climatic changes, and pioneering work by Willi Dansgaard of the University of Copenhagen gave some of the first hints of the speed of climatic change. (See Chapter 6).

A long core of ice drilled from the glacier can be sliced up to be analysed and the temperature variations deduced. A few years ago, Dansgaard and his colleagues reported that this kind of analysis showed that around 90,000 years ago the Earth suddenly plunged from a climatic state rather like that of today into a full ice age, in less than 100 years. Then, however, conditions returned back to the warmer situation within a further 1,000 years. This is, to say the least, peculiar. But the same sudden cooling and rapid recovery shows up in other kinds of climatic records. Fossils gathered from the sea bed near Mexico, changes in the forest cover of northern Greece and the Netherlands, and even studies of slowly growing stalagmites, all show evidence of this event. Such changes may be rare. But the fact that even one has occurred – and relatively recently by geological standards – means that it could happen again. Perhaps a vast series of volcanic eruptions could produce the quick cooling, (see Chapter 8); another idea is that the Antarctic ice sheet could suddenly surge forward, with ice spreading over the southern ocean and cooling the world by reflecting away the Sun's heat. But if volcanoes had caused the drop in temperature, they would surely have left traces of their blanketing dust veil in the sediments, and none are found at the right time. Equally, a surge of the Antarctic ice, although not impossible, ought to have caused a tidal wave which would temporarily raise the sea level by perhaps 200 feet, and that too ought to have left traces in the geological record. But there is no such evidence; the sudden cooling of 90,000 years ago remains an enigma.

That event may have been something out of the ordinary. But it is now quite clear that ice ages are far from unusual in the recent history of the Earth. Probably the underlying reason for the Earth's present tendency towards iciness is the present distribution of the continents, with most land surface concentrated in the northern hemisphere and surrounding a nearly landlocked Arctic Ocean, as we saw in Chapter 8. A few million years ago, with the continents arranged differently, the response of the weather to changes in outside influence could have have been quite different. But that is certainly not something we need worry about in practical terms.

Throughout the first half of this century geologists were content with the idea that there had been four great ice ages, each lasting about 100,000 years with intervals of warmth between them lasting for between 125,000 and 275,000 years. That was certainly a comforting picture, since the most recent ice age ended about 10,000 years ago; according to this view we could expect another 100,000 years at least free from the threat of ice. But the new picture changes all that. In 1955 Italian-born Cesare Emiliani presented evidence, from studies of fossils from the Caribbean, that there had been at least 7 cold periods in the past few hundred thousand years. Emiliani found this evidence by studies of heavy oxygen, similar to the way in which Dansgaard and others analyse the ice records. Tiny sea creatures take up oxygen in their shells, and this is preserved like the oxygen locked up in ice. The result is that we have a record of how the proportion of oxygen-18 has varied, and that reveals how the temperature has varied. It took well over ten years, however, for Emiliani's breakthrough to be fully recognized, and for his work to be corroborated by other studies on land.

The land evidence came from studies made by the Czech scientist George Kukla, whom we have already encountered as a student of the changing ice cover of the northern hemisphere (See Chapter 4). During most ice ages, the region that is now Czechoslovakia was a cold desert, swept by dry winds which laid down a layer of fine yellow dust. During the 1960s, Kukla and his colleagues studied evidence from more than 30 sites where layers of sediment were exposed through the action of rivers or of man, and found several layers of this dust. They also checked in the sediments for pollen and traces of the shells of species of snails which dislike the cold. The result was conclusive and intriguing: at least 10 ice ages, with apparently a regular cycle of cooling and warming.

By the end of the 1960s geologists had a new tool with which they could date these changes more precisely. A magnetic reversal (see Chapter 7) occurred 700,000 years ago, and left its traces in sediments on land and under the oceans around the world. Now, we know that 8 cycles of cooling and warming have occurred in the past 700,000 years, and that a further 12 or more occurred between 2 million and 700,000 years ago. The pattern of these cycles is for warm periods only to last for about 10,000 years or so; according to the new picture we are just about due for a full ice age to set in.

Some people explain the pattern by a slow variation in the Sun's output every 100,000 years or so. Nigel Calder, who wrote *The Weather Machine*, prefers another theory: that variations in the orbit of the Earth as it wobbles around the Sun cause changes in the amount of heat reaching the Earth's surface. These rhythms certainly have the right kind of variation, over the right kind of time-scale. But I am a little less sanguine than Calder about their value for predicting future changes in *detail*, rather than explaining past changes in broader outline. The idea was originally put forward by a

Yugoslav geophysicist called Milutin Milankovitch some 50 years ago. But since then, and until very recently, there was no hope of making his theory fit the standard ice-age pattern, because that pattern of just 4 major ice ages was itself hopelessly wrong. When he was preparing his book, Calder updated Milankovitch's calculation – with results that might even have surprised Milankovitch.

At present, the Earth is tilted at 23.4° relative to the plane of its orbit around the Sun. In July, the North Pole nods towards the Sun, and in January it is the South Pole's turn for warmth. But the Earth's orbit around the Sun is not exactly circular, and reaches 3 million miles further out in July than in January. So summer in the northern hemisphere is not quite as warm as summer in the southern hemisphere. The difference today amounts to some 7 per cent in terms of heat received from the Sun – but the pattern is not always like this.

The Milankovitch theory depends on just how that pattern varies. First, the Earth's orbit around the Sun can 'stretch' from being almost circular to slightly elliptical and back again (see Fig. 17). This stretching cycle takes somewhere between 90,000 and 100,000 years, although the amount of stretch and the exact time taken for a cycle vary a little. The most extreme effect produced by this process reduces the intensity of sunshine reaching the Earth by 30 per cent; but just now we are moving back towards a more nearly circular orbit.

Another possible influence on climate is whether it is the northern or southern hemisphere that receives the benefit of the closest approach to the Sun. At present, southern summers are warmer; but 10,000 years ago it was the northern hemisphere that had the hotter summers. Today this wobble of the Earth (rather like the wobble of a spinning top) has brought us to the worst situation, in terms of northern summer sunshine, that this 21,000-year cycle produces.

The third and last contributor to the Milankovitch effect is a 'roll' of the Earth which varies the tilt of the axis between 21.8° and 24.4°. This roll repeats every 40,000 years; the greater the tilt, the more pronounced the difference between summer and winter. For the past 10,000 years the tilt has been getting less, which should produce cooler summers and warmer winters, excluding other factors.

Clearly the various contributions of these effects are going to cause a fairly complex variation in the amount of heat reaching the northern hemisphere in summer. And that variation is the key to the Milankovitch effect. Because most of the Earth's land area is in the North, the northern hemisphere may be the key to ice ages. Snow and ice cover build up each winter on land, as we saw in Chapter 4, and if the northern summer were a little colder perhaps it would never melt before the next winter came, which would rapidly lower the temperature.

The interactions of the three parts of the Milankovitch effect combine to produce a pattern which is not simple, but does repeat with roughly the same regularity as past ice ages. Calder has done the calculations, and produces the comparison between Milankovitch theory and ice-age records which is shown in Figure 18. As he puts it, the two graphs 'bear more than a passing resemblance' to each other, 'with the same sorts of peaks and troughs at the same intervals'. But as he goes on to say, 'difficulties come with the details — here, perhaps, intense sunshine apparently coinciding with deep ice ages; there low sunshine in periods reckoned to be warm.'* This kind of comparison of two jagged graphs is very subjective; some people see

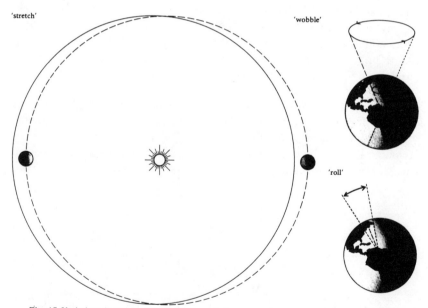

Fig. 17 Variations in the Earth's orientation relative to the Sun, and the 'stretch' of its orbit around the Sun, can change the amount of heat received by the northern hemisphere in summer. (Nigel Calder and B.B.C. Publications)

*The Weather Machine, op. cit., p.133.

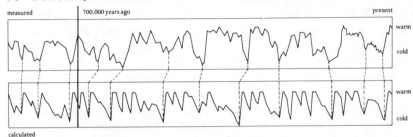

Fig. 18 The 'measured' record of past ice ages (top) is compared with 'predicted' changes in northern hemisphere summer sunshine based on calculations of the Milankovitch effect. The 'measured' record depends on interpretation of oxygen isotope ratios in marine fossils; the 'predicted' record comes from Calder's calculations, made assuming that whenever summer sunshine above 50° N is 2 per cent stronger than at present ice cover diminishes and when sunshine is less ice accumulates. In each case, the rate at which ice cover changes depends on the exact amount of summer sunshine. (Nigel Calder and B.B.C. Publications)

correlations where others do not. I leave you to judge for yourself how much faith you put in the resemblance between these two particular graphs; to my eye, it seems that the Milankovitch theory, as updated by Calder, goes a long way towards filling in the broad outline of climatic change. But some pretty important 'details' remain to be explained, and the obvious explanation is that other powerful forces are also at work to change the worldwide weather pattern from time to time.

We may well be at the end of a warm period, or interglacial. Indeed, an expanded view of part of Figure 18, shown in Figure 19, suggests that the 'next' ice age began perhaps as much as 5,000 years ago when a phase of long-term cooling began. But at least the downward slope of the predicted curve, based on the Milankovitch theory, is not too steep for a few thousand

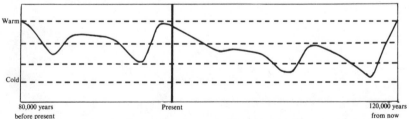

Fig. 19 The calculations used for the 'predicted' curve of Figure 8 produce this more detailed picture of the way the Milankovitch effect has influenced climate in the past 80,000 years, together with a projection of the effect over the next 120,000 years. It seems that we are now 'over the top' of the most recent warm period, and that the Earth is headed for an icy 100,000 years or so. (Nigel Calder and B.B.C. Publications)

years. A lot of ice, lasting about 120,000 years, is the forecast; but it may take a couple of thousand years for the ice to set in.

Calder argues, on the other hand, that past records show a virtual certainty of the next ice age becoming established within the next 2,000 years. However, variations on that short time-scale must be influenced by factors outside the Milankovitch theory – according to at least one study of the patterns of climatic variation (using an analysis of tree-ring thicknesses) after the turn of this century we may be due for 200 years of comparative warmth! So the prospect of a full ice age within 2,000 years need not necessarily cause panic. There must be some prospect, of course, that a full ice age could start even next year. But, as Figure 18 shows, the Milankovitch theory could equally indicate an ice age even if the Earth was enjoying a period of warmth; several periods which ought if that theory is perfect to have been cold were in fact warm, and several 'expected' warm periods were in fact cold.

The variations of the world's climate involve more complex factors than a simple dependence on the variations in northern summer heat produced by the variations in the Earth's position as it rolls around the Sun. But even a small chance (50 to 1? 100 to 1?) that a new ice age is 'just around the corner' cannot be ignored; if only as an insurance policy we should have some idea of what might happen, and we should be prepared, on a world scale, to meet the prospect as best we can. Such preparation must take decades at least, and I believe that we have decades at least in which to prepare. So, in the spirit of providing a clue about what *might* happen within our lifetimes (but which I believe probably will not happen), here is an outline of what Calder has termed 'the snowblitz'.

This idea, that ice and snow cover of the land in the northern part of the northern hemisphere might just fail to go away after one winter, was first developed by Hubert Lamb and a colleague, Alastair Woodroffe, in 1970. One possible scenario for the onset of a snowblitz is that volcanic dust obscures the Sun for a year or two; another is that the atmospheric circulation in the northern hemisphere might stay set in a pattern which encouraged the snow for long enough for the snowfields to become established. If that happened, the very presence of the snow and ice cover would encourage that circulation pattern to.persist. The exact trigger for such a change remains speculative – perhaps some hiccup in the Sun? Whatever the immediate cause might be, a look at the latest discoveries about glaciation to the south and west of Britain shows how quickly things can develop.

The old ideas suggested that ice spread slowly down from the north, taking thousands, or tens of thousands, of years to reach the latitude of the English Channel, if it ever got that far south at all. But in 1969 some of these ideas began to look a little shaky, when earth-moving operations for a new motorway in Somerset, in the west of England, revealed the debris laid down by a great glacier. Geologists could tell from the distribution of this debris that the glacier had flowed not from north to south, but from west to east; that is, from the direction of the shallow sea south of Ireland, the region known as the Western Approaches.

How could this be? With the new evidence to prompt them, geologists took a fresh look at other evidence, and in 1974 proposed that the English Channel itself is the site of a former glacier, fed from a great dome of ice in the Western Approaches area. And the explanation of how such a glacier could grow up ties in exactly with the snowblitz idea. What now seems to have happened is that, as the Earth began to cool in an ice age, enough water was frozen to reduce the sea level to such an extent that the shallow sea south of Ireland dried out. There, the prevailing westerly winds off the Atlantic, bringing their burden of moisture from the sea, dumped great quantities of snow. The new snowfield rapidly became a great ice dome, building up until glaciers began to slide away off it, some to produce icebergs in the Atlantic, others to move eastwards across the southwest tip of England and up the English Channel. Geoffrey Kellaway, a geologist who has pioneered this new concept of ice ages, suggests that similar ice domes would build up to the north of Ireland, and in the present North Sea between Scotland and Norway.

This evidence completely contradicts the old idea of a slow spread of ice from the north; it will be interesting, over the next few years, to see how geologists' understanding of ice movements in other parts of the world are modified by these latest discoveries about the spread of ice on the eastern margin of the Atlantic. With such evidence, combined with the evidence of the ice cores, pollen studies, and analysis of the oxygen isotopes found in the shells of tiny marine creatures, the snowblitz idea must now be accepted as the best explanation yet for the onset of an ice age. But if that theory is right, a snowblitz may not occur in the next hundred years. It's a measure of how much man's ideas about ice ages have changed since the end of the 1960s that I would not be prepared to say that such an occurrence is impossible. Indeed, it's *possible* (although extremely unlikely) that by the time you are reading these words winter snows in the northern hemisphere will already be persisting beyond their expected duration, and that we will be in the beginning of a snowblitz. In my view, that is as likely as the rather remote possibility that some unexpected change in the workings of the Sun will heat our planet unendurably in the near future. To put things further into perspective, the chance of nuclear war breaking out on Earth is rather greater than either of those two possibilities.

As far as climatic change and the threat of ice are concerned, there are three bodies of opinion: the pessimists say that climatic doom is imminent, and if they are right then there is so little we can do that it hardly seems worth trying. One might as well crawl into a cave and wait for the end. At the other extreme, there are still a few super-optimists, whose outlook is not unlike that of believers in the Flat Earth Theory. They say that climate doesn't change very much in any time-scale relevant to man and that there is nothing to worry about.

The third group I would call optimists, and I number myself among them. Their view is that an ice age may be upon the world within a few hundred years, and that the immediate deterioration of the climate at the present time requires urgent attention from all responsible people. That may only seem optimistic in comparison with the pessimists' view; but I feel genuinely optimistic that if we can get over the problems facing us in the next couple of decades, then within a hundred years or so we may well be in a position to adapt our global society to withstand even the rigours of a full ice age. When you look at the related problems of overpopulation, lack of food, small-scale climatic change (let alone the snowblitz) and shortage of energy it certainly does take an optimist to believe that we will get through those crucial decades which will take us up to the end of the twentieth century. I shall look at how the more likely changes will force us to modify our habits in Chapter 12; but first it's important to know just how far ahead we can predict the weather accurately, and how well our growing understanding of the workings of the weather will allow us to plan the necessary changes in the way we live.

11 Conventional Forecasting Techniques – and Their Limitations

Studies of the history of climate, of the solar and other cycles, and the Milankovitch theory, all constitute one approach to climatic change. The other approach is to calculate how the atmosphere should respond to changes, and use these calculations to produce exact mathematical forecasts of how it will evolve – in other words, how the weather will change – starting from the present-day situation. The problem of such mathematical modelling is a huge one, even with the aid of modern high-speed electronic computers. Even predicting the weather for a day or two ahead for a limited area of the globe is pushing the numerical approach to its limits at present. There are so many calculations involved in predicting the growth of a storm, for example, that it takes nearly as long to do the calculations as it does for the real storm to evolve. (Indeed, only a few years ago it took longer!) At present, the best weather forecasts are obtained by combining the computer modelling with conventional studies of meteorological measurements by experienced meteorologists. It might seem, from the trouble mathematical modelling has in dealing with local weather, that its use in climatic work would be even more restricted. But that is not quite true.

The best computer models of the whole atmosphere do work in the sense that they produce the right kind of atmospheric circulation, with the trade winds, circumpolar winds, and so on. But the computer can only be fed with a limited amount of information. In one model used by researchers at the U.K. Meteorological Office, the atmosphere model is represented by 'measurements' of temperature, and so on, at 10,000 grid points around the globe, for each of 11 levels in the lower atmosphere. That means the computer is working with a three-dimensional 'grid' of 110,000 points. This model produces a very good pattern of the resulting prevailing winds, including the general outline of the monsoon winds which are so important for the Indian sub-continent. But with 10,000 grid points at each level, that allows just six points at each level to cover the whole of that sub-continent. The scale of the model is far too large to pick out the relatively tiny variations in monsoon winds that have been responsible for recent droughts, floods and famines in the area.

So we are still a long way from using this kind of mathematical modelling to predict details of climatic change. That is not say, however, that those techniques are not of the greatest importance to climatologists struggling to develop a thorough understanding of the workings of the atmosphere. The reason is that when meteorologists are studying the broader canvas of climatic variations they can leave out a lot of the details that are essential in standard weather forecasting. Even the fact that the models can show the main features of the differences between global circulation in January and July is a great step forward from our knowledge a few years ago. For the particular Met. Office model I have mentioned, the computer takes about 10 minutes to 'evolve' the atmosphere through one model 'day'; but the wind-speed numbers fed into the computer must be fairly sensible or it will produce a nonsensical pattern of atmospheric circulation.

The next step in the development of these numerical models is to include the effects of cloud cover, allowing the models to 'make' clouds. This is particularly important in climatic change, as we have already seen. Clouds are a vital factor in adjusting the energy balance between radiation absorbed from the Sun and heat radiated out again from the Earth into space. As yet, the mathematical modellers are still at the stage of developing models which can explain all the details of the *present* climate; only when these are complete will they be able to move on with confidence to predicting how certain changes in the heat balance will cause future climatic change. But in some limited areas, computer simulations can provide valuable information about how detailed changes in the atmosphere, or snow and ice cover, are likely to affect climate. Unfortunately, a lot of the time the 'answers' are either ambiguous or confusing.

Studies at N.C.A.R. in Boulder, for example, show how difficult it is to predict the likely effects even of increased cloud cover. It has long been taken for granted that more clouds mean more reflection, and therefore lead to a reduction in temperature. But increased cloud cover is caused simply by the presence of more water vapour in the atmosphere, and the resulting effect on temperatures seems now to be closely linked to the *kind* of clouds formed by the extra water vapour.

If we start out from a typical situation with fluffy, medium-sized clouds and work out how cloud changes will affect the temperature some interesting answers emerge. Taller clouds, which block out very little extra sunlight, also have colder tops, and the colder tops radiate less heat than the warm tops of short fat clouds. So a shift in cloud cover which produces more tall, thin clouds can lead to an *increase* in global temperatures, not a decrease. A spread of wide, low clouds probably will cause global cooling, because they block out a lot of heat, and they are also warm enough to radiate a lot of heat into space. But the numerical models used in this N.C.A.R. study show that we are still far from a thorough understanding of how changes in the present pattern of global weather conditions will lead to long-term climatic changes (see Fig. 20).

Another 'answer' to a hypothetical climatic problem, this time one produced by the Met. Office computers, also caused some rethinking among climatologists in the early 1970s. This time, the computers were asked to calculate the same kind of general atmospheric circulation model as usual, but leaving out the effects of the Arctic ice cap. The mathematical models use information about meteorological parameters at the grid points to calculate circulation patterns, and then use the effects of these theoretical winds to update the numbers corresponding to temperatures, pressure, and so on, at the grid points in a continuously evolving cycle. So it is straightforward to pull out of the computation an estimate of the effect on temperatures in the northern hemisphere of the changed circulation resulting from removal of the ice cap.

Surely, you might say, the absence of ice cover corresponds to a warmer state of the Earth? If that's what you would expect, the result of these particular calculations will come as a surprise. The result of the changed circulation, at least in the models so far studied at the Met. Office, is to *reduce* the average temperature over Europe by about 2°C. A rise in temperature can be produced, but only by removing a small fraction of the ice cap and leaving the rest here. Why should that be the case? When the whole of the Arctic ice is removed the entire circulation pattern of the northern hemisphere is changed; without an ice cap, the westerlies, which dominate circulation at mid-latitudes, are displaced to the south, and regions which at present benefit climatically from the influence of the westerlies become cooler.

But even these numerical model calculations must be treated cautiously. Remember that, among other things, the models used in this study did not include the effects of cloud cover. In the real world, if the Arctic ice ever did

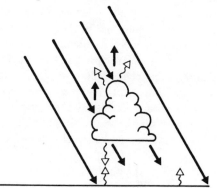

Fig. 20 Effect of different kinds of cloud cover on heat balance of the Earth.
(a) An average cloud reflects about 50 per cent of solar radiation, and also radiates (wavy arrows) some of the heat it absorbs.

(b) A low, wide cloud reflects more heat – but also blocks radiation from the surface below. On balance, when reflection of sunlight dominates the Earth cools.

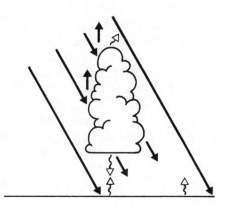

(c) A tall, thin cloud reflects as much as the average cloud in case (a) – but because its top is cooler, it radiates less heat. The result is a net warming of the Earth. (N.C.A.R. *Quarterly*, February 1974)

disappear then there might well be a lot more water vapour in the atmosphere. Would that produce tall, thin clouds to warm up Europe? Or would broad, low clouds encourage cooling to go still further, hastening the return of the ice? We just don't know. The question of clouds is probably the greatest bugbear of all. Certainly clouds are of the greatest importance. One surprising result that has come out of satellite studies is that the amount of heat radiated by the Earth is exactly the same in the northern hemisphere as in the south, although the southern hemisphere contains very little land. It can only be the clouds of the southern skies which account for this.

And what of the oceans? The climate depends very much on the complex interactions of atmosphere, oceans and ice together, and these interactions must be understood if we are to interpret properly what has gone 'wrong' with the weather lately. Currents in the sea produce their own kind of 'weather', and the great warm currents in particular (such as the Gulf Stream) have a great influence on atmospheric circulation. Ocean currents transport heat from the warm tropics to the cooler polar regions – and to a far greater extent than was realized even a few years ago. There are eddies in the sea which seem to work in much the same way as the depressions in the atmosphere; the oceans now seem to be made up of many almost distinct layers, with clear-cut boundaries between them, where abrupt changes in temperature, salinity, and so on, take place.

All these features of the ocean are really only just becoming known, and are far from understood. The subject is so new that many of the theories are conflicting, and most of them (if not all) will soon be out of date as the study of the oceans develops. So I shall only mention in some detail the way in which variations in the warmth of the oceans can be used to help us understand some changes in atmospheric circulation.

The climate of North America depends very much on the exact path of the jet stream, the strong permanent wind high in the atmosphere which blows its zig-zag way around the North Pole from west to east. At the beginning of the 1950s the jet stream brought relatively warm air and mild winters to the northern U.S. But in 1957 the pattern of warm and cold areas of surface water in the Pacific changed dramatically, and with this change came a change in the usual route of the jet stream and harsher winters which persisted through the 1960s. The pattern then established was one in which central Pacific waters north of Hawaii were cold, with warm water established down the western seaboard of the U.S. The jet stream swung south across the cold patch, before turning north around the warmer water and then sweeping almost due east across the continent. In 1971, the pattern of warm and cold water reversed again, and the jet stream swung north around an area of warm water in the central Pacific before diving down south to California and swinging up towards the northeast across the U.S. This gives warmer winters in the eastern U.S., while the western states, alongside the area of cooler ocean water, have colder winters.

These changes in the jet stream are also related to warmer winters in the most highly populated areas of Europe. So it is a striking paradox that while the world as a whole probably is cooling down, the most technologically developed areas of the U.S. and Europe have had a run of mild winters, which naturally makes most ordinary people rather puzzled when confronted by meteorological claims that a new ice age is on the way!

Variations in sea-surface temperatures (S.S.T.s) are used in forecasting weather in 30-day forecasts of the U.K. Met. Office, though it must be admitted that these forecasts seem to produce results which are only just better than guesswork. But this work provides clues which are obviously related to the evidence we have of how the S.S.T. variations of the Pacific affect the weather of North America. It's too early yet for these clues to be incorporated in a workable complete theory of the interactions of ocean and atmosphere. But at least they provide a starting place for the theorists.

What the Met. Office studies have shown is that in forecasting the circulation and weather around Britain for a month ahead, it is important to know whether the Atlantic Ocean south of Newfoundland is warmer or colder than usual. This fact emerges from study of records going back to 1880; since the combination of good weather information for Britain and S.S.T. measurements south of Newfoundland is just about unique in going back that far, there is much less evidence about how Pacific S.S.T. anomalies affect the weather of the U.S. from month to month, although there does seem to be a similar kind of effect.

The ocean temperature variations are produced as charts on which S.S.T. variations (or 'anomalies') are marked in areas covering 5° in latitude and 5° in longitude. The measurements come from a wealth of sources, such as British and U.S. naval records, and it took a good deal of detective work to sort them all out. But when they were sorted out a clear pattern emerged. Most of the S.S.T. patterns in the area south of Newfoundland since 1880 fall into one of eight classifications. Three of these are warm sea types, three cold, one for sea cold in the southern part of the region and warm in the north, and the other for the opposite case. The details of each type of S.S.T. distribution are not important unless you are working on the forecasts; what is important is that each type seems to correspond to a certain weather pattern over Europe one month later.

So the Met. Office has used this kind of evidence, as one factor among many, in preparing the 30-day forecasts that it issues each month for the U.K. The S.S.T. forecasts can, of course, be unravelled from the overall forecast and tested against the actual weather, and it turns out that if the S.S.T. evidence alone had been used, during the three years 1969-72, 75 per cent of the temperature forecasts made on this basis alone, and 66 per cent of the corresponding rainfall forecasts, would have been, as the Met. Office puts it, 'in moderate agreement' with actual events. By 'moderate agreement' they mean better than a chance agreement. Now, this is not a great forecasting breakthrough. since any intelligent observer of the weather in Britain could probably also produce better than chance forecasts, simply from a knowledge of the general behaviour of the weather and a specific knowledge of what the weather has been doing in recent months. But if the S.S.T. distribution south of Newfoundland is not an exact guide to the weather for the month ahead in Europe, the fact that the two are related at all does provide another vital clue as to the workings of the ocean/atmosphere system. An even better clue is provided by the correlation for the winter months alone between the S.S.T. patterns and the weather − about 90 per cent of these forecasts are in 'moderate agreement' or better with actual events.

Clearly, ocean temperatures have a bigger effect on the weather in winter than in summer − and this ties in well with what little we know about the workings of the oceans. As the summer develops each year, a layer of warm water builds up at the top of the sea, eventually extending down to about 30 metres. At the bottom of this layer, a very distinct boundary (the 'thermocline') separates it from the deep, cold body of the ocean. For practical purposes, the atmosphere interacts only with the warm surface layer once it is established − and because this is such a relatively thin layer it is subject to erratic fluctuations in temperature.

In winter, however, the thermocline is essentially at the top of the ocean, and the whole depth of water down to the sea bed acts as a 'constant temperature bath' influencing the bottom layers of the atmosphere. That is a more stable situation, producing longer-term effects that are reflected in the weather to the east not just over 30 days but, in fact, throughout the whole winter.

There is some evidence that this forecasting technique would have given a fair guide to the winter of 1974-5. But the limited success of that 'forecast' shows the difficulty of trying to relate global (or hemispherical) averages to events in limited regions of the world. According to the S.S.T. situation in late 1974, it seemed that Britain and Europe could expect a severe winter. For most of Europe, that prediction was borne out, but in Britain we experienced, for the fourth successive year, a distinctly mild winter. What went wrong with the forecast? In global terms, the British Isles are very small, and it seems that the U.K. was caught in a stream of relatively mild, wet prevailing winds from the Atlantic. This was a local irregularity, as far as the northern hemisphere was concerned, related to the pattern of circulation of the jet stream and the circumpolar lows that became established with the weak circulation pattern. Similar ripples in the circulation pattern brought mild winters to a handful of other small areas of the northern hemisphere. But, taking the hemisphere as a whole, the trend of average temperatures is still slightly downward. The trouble is, at present we can only explain the anomalies with hindsight — we can't yet predict just where these local ripples are going to show themselves.

It does seem that in the grey area between straightforward weather forecasting (over a few days, or perhaps a couple of weeks) and genuine climatic forecasting (over periods of several years or more) the greatest use of conventional forecasting techniques may well be in seasonal forecasting. We are a long way yet from accepting any kind of seasonal forecast; but with a combination of the speed of electronic computers, the intuition of experienced meteorologists and a developing understanding of air/sea interactions, seasonal forecasting is something more than a dim prospect.

I am less optimistic about the success of conventional forecasting techniques applied to fundamental climatic change, and the likelihood of achieving success in predicting such changes through numerical modelling in particular. There are those who argue persuasively that only exact mathematical models can be trusted to make predictions on which to base the actions of nations, and that computers will eventually develop to the point where they can handle the vast amount of information about atmosphere, oceans and ice needed to produce reliable models on which forecasts can be based. But I see little chance of this, certainly for a long time ahead.

Chuck Leith, of N.C.A.R., is among those who doubt that such reliable numerical modelling of climate is attainable even in principle, regardless of practical difficulties. On any numerical simulation of a real problem of this kind, it is a well-known rule of statistics that fluctuations with a size inversely proportional to the square root of the number of samples can be introduced by 'sampling error' alone. In practical terms, if we make 4 measurements of daily temperature, say, and try to work out what the next day's temperature will be, then we can expect an error of up to half (= root (1/4) if the day-today variation in temperature. Suppose our four measurements of temperature were 9°, 10,° 13° and 10°C respectively. Then the 'average' temperature is $(9 + 10 + 13 + 10) \div 4 = 42 \div 4 = 10\frac{1}{2}$°C. But it would be quite misleading to quote that precise figure as a reliable average. With only four samples, we can expect an error of up to $\frac{1}{2}$ of the spread of temperatures measured. The spread is 4°C $(13 - 9)$, so we expect an error of up to 2°C. Our 'average' should be quoted as $10\frac{1}{2}$° plus or minus 2°C, which is much less impressive. But if we carried on our measurements for a total of 16 days and found the same spread (between 9 and 13°C) and the same average temperature, we could have much more confidence in the reliability of the average. Now, the expected error is 4 divided by the square root of 16, which is also 4. So the error is one degree, and we can quote the average as $10\frac{1}{2}$° plus or minus 1°C. This immediately tells us that the days on which the temperature reached 13°C were unusual – a full $2\frac{1}{2}$° above the average, and, more significantly still, a full $1\frac{1}{2}$° above the (average plus expected error) temperature, which is $11\frac{1}{2}$°C. In principle, the accuracy of the average can be improved indefinitely by taking more and more measurements. But this is tedious and time-consuming, and in practice climatologists and meteorologists must live with the fact that there is always an unwelcome statistical fluctuation, or error, of this kind built in to their figures.

There must be fluctuations in the atmosphere, which cause a cloud to drop its moisture as rainfall somewhere within a wide area. There must also be statistical fluctuations in the numerical models, and the exact size of these fluctuations depends on how big the mesh of the worldwide 'grid' of input information is (as we saw, even 10,000 grid points provides only a crude guide to the workings of the atmosphere) and on how often the input data are being updated as the numerical model is 'evolved' by the computer. This may make the prospect of accurate numerical forecasting of climatic change seem pretty unlikely; I go along with the views of Hubert Lamb, who has commented that numerical models, while invaluable as an aid in developing a broad understanding of the processes which are involved in changing both weather and climate, simply cannot cope with the complexities and detail which would be required for numerical modelling alone to provide us with accurate forecasts for the seasons, years and decades ahead.

The best hope of understanding why the weather changes, and of predicting what is ahead, seems to me to lie in the synthesis of the different approaches I have discussed. Numerical modelling helps to provide new insights – such as the surprising possibility that less ice in the north might lead to cooler climatic conditions in Europe; we also need a better physical understanding of variations in the heat balance of the Earth, whether it be volcanic effects, the Milankovitch process, or some other cause that produces the variations; and as we have seen, study of patterns of historical variations helps to provide the clues to the underlying physical causes.

Climatic forecasting is still partly a process of informed guesswork and partly a process of inspiration; but we can say something about the most likely pattern of climatic change over the next few years. Bearing in mind the present limitations of all such forecasts, at last we are in a position to sum up the best informed view of what's likely to happen to the weather next, and how it will affect our daily lives.

12 Climate and World's Food Supply Over the Next Few Years

Changes in climate can affect much more than our food supply, especially when we start talking about the kind of big changes that might be associated with a new ice age. Even on a smaller scale, it has been estimated that a fall of 1°C in the average temperature of Britain would bring an increase of 2 per cent in the demand for energy. But agriculture is probably the area most sensitive of all to changes in the climate – we have already seen the dramatic repercussions of a small shift in the global circulation pattern. This is also the area where such changes cause real hardship, not just inconvenience. So it is natural to concentrate on the relation between climate and the world's food as the area of prime concern.

Any intelligent person living in an industrialized society can see how a small change in temperature would affect that society. More fuel for heating, transport difficulties, even such seemingly mundane problems as burst pipes can add up in a terrifying manner, as the cold winters of 1947 and 1963 showed. The world economy would also be affected by climatic deterioration through the demand for more efficient agriculture. More irrigation, more and better fertilizers, more tractors, and so on – all required to make farming more efficient to cope with the problems of a shorter growing season, droughts, and the like. This change in agriculture patterns would by no means be insignificant. To take just one example, it is ironical that India, for political reasons, has been a supporter of the Arab nations' policy of restricting the availability of oil and increasing its price. Yet oil is not just used to lubricate the wheels of western economies; it is also required in the manufacture of fertilizers. So this action, which was supported by India, has led directly to there being less fertilizer available for Indian farmers, at a time when that country has a serious food problem. It has been pointed out that the U.S. and Canada between them control a larger fraction of the world's exportable food reserves than the fraction of oil reserves controlled by Arab states.

What exactly are the implications of a declining climate on food? This is probably the biggest problem facing mankind today; and if we do not solve the first and biggest problem there is no point in worrying about the others. Even the energy crises must be secondary in importance. First of all, before we look at the problem of how the changing climatic pattern is likely to affect food supplies, just how is the climate likely to change between now and, say, the end of the twentieth century? With all the evidence gathered in this book, we are in a pretty good posidion to make some realistic forecasts – informed guesses, if you prefer – without getting carried away by the possibility of imminent disaster.

The first thing we can do is to rule out any influence of man's activities on the climate. As we have seen, these activities might be producing a slight net cooling, or a slight net warming, or no effect at all (see Fig. 16, p. 89) I think that the picture of man-made dust rushing us into a new ice age is what the economic doomsdayers call a 'worst possible scenario'. I am not saying that man should continue polluting the atmosphere regardless, of course. Just because we cannot prove that something is harmful does not mean that it is harmless. The worst hazard seems to be the threat to the ozone layer posed by indiscriminate release of the gases used as propellants in spray cans, and the release of exhaust gases from high-flying supersonic aircraft. I still think that the ozone layer is probably a pretty stable feature in the upper atmosphere, and that it won't be easily destroyed. But I don't think we should take chances, and I would like to see a complete ban on the use of these spray cans – and on supersonic aircraft like Concorde – purely and simply as an insurance policy. In the same way, we should be making every effort to minimize the emission of smoke from factories, and the destruction of natural forests, which play a part in determining climatic patterns. With that cautious proviso, however, for the present practical purposes we can surely neglect man's direct influence on the climate.

So what are the hazards of natural climatic change? They seem to fall into two categories: those produced by events in or on the Earth; and those produced by changes in the heat we receive from the Sun. The former remains something of a mystery. Perhaps volcanic dust plays a part in causing global cooling – especially after a dramatic explosion like that of Krakatoa – but it is difficult to see how this can explain the regular cycles of cooling and warming that the Earth seems to undergo (see Fig, 18, p. 96). We can't really plan our agricultural activities on the assumption that another Krakatoa may explode next year. The same is more or less true of other natural effects on Earth. If the Antarctic ice suddenly fractured, sending a surge of ice north, there would be little or nothing we could do about it.

In fact, the only way we can make any worthwhile plans to cope with the changing climate is by finding patterns of climatic change which enable us to make reasonable predictions. The possibility that climatic patterns drift westward as the Earth's magnetic field drifts westward would be helpful if the theory were thoroughly proved; but until a lot more research has been done it cannot be a basis for agricultural policy. Variations in the Earth's orbit, as suggested by Milankovitch and modified by Calder, provide a very good long-term guide. In my view, that leaves us with two related possibilities for short-term predictions. First, there is analysis of past climatic trends – over periods of tens, hundreds and a few thousands of years – which, even though not wholly understood, can be useful for forecasting. Second, there are the natural variations of the Sun which occur ov the sunspot cycle, and the variations in the strength of the sunspot cycle itself, which seem to occur over periods of about 90 and 180 years. Sunspot variations can be accepted empirically, leaving aside the question of their cause, in the same way as climatic cycles. And when climatic and solar cycles coincide our faith in this empirical approach is greatly strengthened.

Where does that leave us? To start with, we seem to be experiencing a global cooling trend. In fact, the evidence for this is not conclusive; Gordon MacKay, Director of the Meteorological Applications Branch of Environment Canada, told me on a visit there in the autumn of 1974 that some recording stations in Canada have shown an increase in temperature over the past few years, although the trend of temperature in the great plains (all-important for wheat production) is indeed downward. As Stephen Schneider of N.C.A.R. put it when I spoke to him, the meteorological observations 'are inadequate to support the conclusion' that there is a *worldwide* decline in temperatures; it remains possible that this widely publicized trend is artificial due to the locations of the relatively few sites around the globe whose temperature figures are averaged to give the worldwide trend. But we can't afford to take a chance on this apparent cooling trend not being real. So the sensible approach seems to be to expect the overall temperature trend to continue downwards over the next 25 years, returning us to conditions rather like those at the beginning of this century. With temperature changes and the spread of ice, as we have seen, there are associated circulation changes affecting rainfall, with rainfall belts continuing their movement southwards in the northern hemisphere and, even more worrying, increased rainfall occurring over the oceans and less over land (see Fig. 5, p. 16).

Agriculture is likely to suffer two main effects: a shorter growing season and less rainfall. The real urgency of the present situation lies not in any extravagant fear that Britain, Europe and North America will be covered in ice within a century, but in the more mundane (but more likely) prospect that climatic influences will make it harder to grow food just at a time when the present population has increased to the point where food supplies are precarious (see Fig. 21). Ironically, at the beginning of the 1970s there were not only reasonable reserves of food but also reserves of idle agricultural land, notably in the U.S., which could have been used to build up the world's emergency reserves. Instead, the U.S. and Canadian Governments paid many farmers a subsidy to stop them farming their land and producing grain which could have caused a decline in prices and a 'bad' effect on the economy. As the world population and demand for food has grown, and the climatic conditions for growing food have slightly deteriorated, all of this land has been pressed back into agricultural use, while the reserves have

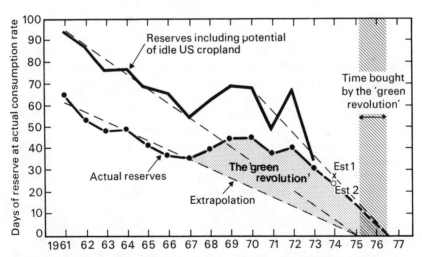

Fig. 21 World food grain reserves for each year since 1961. The steady decline in reserves was temporarily offset as idle land was brought into use, and by the effects of the 'Green Revolution'. That revolution is now over, and there is no more spare land. The extrapolations on this figure (provided by Reid Bryson, using data gathered by American climatologist Lester Brown) take no account of the effect of any continuing deterioration of the climate.

been run down. At the same time, the success of the 'Green Revolution' of the middle and late 1960s, in which yields per acre of many crops were dramatically improved by the use of chemical fertilizers and better farming methods, seems to have run its course; we can expect no more spectacular results from science in terms of better yields from poorer soils.

As Figure 21 indicates, the grain supplies from North America are all-important in such a situation. Because the U.S. and Canada provide so much of the exportable grains in the world, a small decline in their output produces a big effect worldwide. A run of bad years might cause a 5 per cent decline in agricultural productivity in North America – not enough to worry the people who live there, but enough to knock 20 or 30 per cent off the amount of grain available for countries which can't grow enough of their own. And the predictions of Figure 21 now look a little optimistic; in the fall of 1975 the failure of harvests in the U.S.S.R. to reach expectations brought Russia back into the grain market as a big buyer at a time of renewed shortages. For the past 10 or 15 years, a combination of improved technology and a reasonable run of weather has maintained at least something of a lead of supply over demand. We now need good weather for several consecutive years simply to get back to the position of comparative safety that was abandoned at the beginning of the 1970s.

Is there any evidence that, whatever the longer-term trend, we might be due for a lucky break in the form of four or five years of more amenable weather conditions and better harvests? For once, there is just a ray of hope. The sunspot cycle, which bears a definite relationship with climate (see Fig. 9, p. 41) reached a minimum in the mid-1970s, and is working up to a new peak. Astronomical opinions differ on whether the peak will occur just before 1980 or just after; the planetary tide theory of sunspots predicts a broad but low sunspot peak around 1981-2 (see Fig. 11, p. 46), but not everyone accepts this theory. Since more sunspots do seem to bring better weather, it's certainly worth taking a closer look at the evidence of a link between sunspots and agricultural output.

The 1969 peak of world food reserves amounted to 19 per cent of one year's consumption; by 1974, reserves were only 7 per cent of annual consumption, and variability due to the weather can amount to 10 per cent of annual consumption. This rather disturbing evidence encouraged a group of scientists working at the Appleton Laboratory in England to take investigations of a relation between sunspots and the weather one stage further, and to compile any evidence they could find of a relationship between sunspots and agricultural productivity. The evidence is impressive.

Taking the world wheat production figures for the period 1949 to 1973, wheat production in 1958 was greater than in each of the next five years, and production in 1968 was greater than in any of the next four years. Is it just a coincidence that sunspot maxima occurred in 1957 and 1968? Again, in 1954, a year of sunspot minimum, production of wheat was less than in any of the 2 preceding years or the 2 following years; in North America alone, the crop in 1954 was the smallest of any year for which records are available, and the average crop during the 5 preceding years was 25 per cent greater than the 1954 crop. Figures from many individual countries show the same kind of variation of agricultural production in step with sunspot variations. In Canada, for example, during 1967-9, the years centred on the 1968 sunspot maximum, the average wheat production was 27 per cent greater than during the four years 1970-73. In general, over the northern hemisphere wheat production seems to be significantly enhanced near sunspot maxima and reduced at sunspot minima.

Curiously, however, just the reverse seems to happen in the southern hemisphere. In Argentina, for example, booms in the wheat crop occurred at times of sunspot minimum, with 1954 and 1964 being very good years; in Australia 1957, close to sunspot maximum, was a particularly poor year, and the average production over the preceding eight years was 84 per cent greater than in that year. This shows just how little we understand the workings of the atmosphere, and in particular the influence of solar activity on the weather. But this apparent balance between north and south is no great help, since by far the greater land area of the globe, and by far the greater production of wheat and other food, is concentrated in the northern hemisphere.

In many countries, says the Appleton Laboratory team, the modulation of wheat production associated with the solar cycle seems to be at least 10 per cent; in some parts of the world it may even be greater than 50 per cent. There is a story – long regarded as something of an old wives' tale – that stock market fluctuations and other economic indicators follow the sunspot cycle; the effects on agriculture which have now been discovered could have economic repercussions. Obviously such repercussions would have been much bigger in the past, before industrialization; perhaps Elizabethan England used to experience an 11-year cycle of booms and slumps!

A more reliable speculation is that the 'double' sunspot cycle of 22 years has more influence on agricultural output. There is not really enough evidence to prove this, but there were droughts in some agricultural areas of the U.S. in 1954, a year of sunspot minimum, and lack of rain caused problems with many crops during 1974, more or less in line with the 'prediction' of the 22-year cycle.

Looking over longer time-scales, there are also 90-and 180-year periods over which the strength of the solar cycle seems to vary. As described in Chapter 5, these trends suggest that the climate averaged over successive decades will get worse before picking up again in the early decades of the next century; in terms of the problems in parts of Africa, say, rainfall belts will probably continue to drift south for another 50 years. If there is a cooling trend caused by some other effect to be added to this sunspot effect, then the outlook for the rest of this century is poor despite the beneficial effects of the sunspots over the next five or six years.

The optimistic view, then, of our present situation is that with the aid of the sunspots, a recovery could be achieved, with world food reserves being built up over the next few years to the point where we could cope with another run of years as bad, in agricultural and climatic terms, as those of the early 1970s. Our technology is sufficient for the task. But is there the necessary political will? The tragi-comedy of the World Food Conference held in Rome in November 1974 suggests that there may not be.

Population must be limited – if not by famine, war and disease, then perhaps by an increase in the world standard of living, since such an increase always seems to lead to a declining birthrate. But that could only be achieved in the time available by drastically restructuring our global society. Stephen Schneider has compared the budget required to the defence budget of the U.S., which makes the effort just about possible. But he doubts whether countries around the world will make such a shift of emphasis from expenditure on destruction to expenditure on construction; his prognosis is 'technologically optimistic but politically bleak'. A few days after talking to Schneider, in October 1974, I witnessed, in Toronto, an example of just the sort of parochial narrow-mindedness to which he refers.

The Canadian Broadcasting Company's national news on October 30th carried two items, side by side, which illustrated for the most complacent just how confused the situation is. One item reported on the slaughter of 600 calves by farmers making a protest against government policies – their exact reasons are unimportant. The item included film of a calf being killed by having its throat cut and being tossed (still twitching) into a pit for immediate burial. Naturally, this film drew a strong response from Canadian viewers; some protestors described the film as 'obscene' and many said that it should not have been shown on TV. The next item on the same news bulletin concerned the situation in Bangladesh, with film of starving men, women and children. Nobody, apparently, wrote to or phoned C.B.C. to complain that the showing of such film was obscene; and according to C.B.C. none of those who protested at the obscenity of the first film suggested that the obscenity lay in an economic situation which allowed farmers in Canada to throw food away while people in Bangladesh were starving.

Dr Henry Kissinger has recommended as a target a reserve of food of 60 million tons – more than double present reserves. If that suggested safety margin is to be achieved, who is going to produce the extra food, even granted a little help from the sunspots? If it is not, the indications point to a worldwide depression in the standard of living and increase in malnutrition. Big famine disasters may occur, but a more insidious general increase in malnutrition across the Third World is likely to be more serious. Yet in many small, densely populated countries there is no food problem, as agricultural economist Edgar Owens has pointed out. He cites the particular examples of Egypt and Taiwan, where output of basic food grain averages well over 3,000 pounds per acre – more than most of the so-called 'rich' countries, including the U.S., but produced by intensive farming on plots only 2 or 3 acres in size. According to Owens, if Indian farmers could be organized in such a productive way, India could produce a surplus of food grains – in excess of the country's own needs – equal to twice the world trade in such grains in 1972. Even doubling the agricultural productivity of a few medium-sized countries, such as Mexico and Pakistan, would solve the immediate world food problem – and would still leave their productivity at lower levels than those of Egypt and Taiwan. This should be possible – but the kind of effort involved is best seen by considering the example of China.

The economic policies of many countries would have to be altered drastically to assist poor farmers; agricultural policies would have to encourage intensive use of smaller farms rather than inefficient larger farms. Not many countries are yet ready to tackle these problems in the way that has proved successful in China; but the longer we wait the worse the crisis will be.

Postscript

Dusty embrace of the Milky Way may explain Ice Ages – and the origin of man*

Ice Ages on the Earth, and even the very creation of the planets, may have been caused by the passage of the solar system through dust lanes lining the spiral arms of the Galaxy.

Predicting future climate for the long term is a game for the boldly imaginative. Most recent prognostications of this kind talk gloomily of the imminence of a new Ice Age; but one bold imaginer takes a different view. According to the ideas of Professor W. H. McCrea of Sussex University (*Nature*, vol 255, p. 607) the longest long range forecast yet attempted is also the happiest: set fair for the next 250 million years.

McCrea is an astronomer, and brings an astronomer's grasp of the big time (and big distances) to the problem of terrestrial Ice Ages, which is really something of a local problem. Rather than invoking changes in events on Earth to account for climatic changes of this kind, he looks at the place of the Solar System in our Galaxy. One major snag about invoking volcanic dust or sudden surges of the Antarctic ice cap to explain such phenomena has been that although these mechanisms may work in a 'one off' sense, it is difficult to see how they could produce the apparent periodicities seen in the geological record of ice ages. On the other hand, some astronomical phenomena have the right kind of periodic changes, but until now it has been difficult to see why they should be related to terrestrial climate.

* This article first appeared in *New Scientist*, London (June 26th, 1975), the weekly review of Science and Technology.

In very round terms, there is, says McCrea, geological evidence that periods of ice cover occur roughly at intervals of 250 million years. Each of these Ice Epochs lasts for a few million years, and during them conditions vary between those of extreme glaciation and the milder inter-glacials, when conditions are like those of the present day; but such minor fluctuations on timescales of less than a few million years are of little concern to the development of McCrea's model.

This 'period' between Ice Epochs is just half the time it takes the Solar System to make one orbit around our Milky Way galaxy, and a previous generation of astronomers has already noted this coincidence. Could it be more than a coincidence? If it is, the effect must presumably be related to the spiral structure of our Galaxy, since there are two spiral arms each of which will be crossed by the Solar System once in a complete 500 million year orbit. And that is where McCrea's new work goes a bold step further than the speculations of his predecessors.

Spiral arms are a feature of many galaxies, and we can understand our own Milky Way system by comparing it with others. Along the edges of spiral arms there are dark lanes of dust and gas (see figure 22). These, according to the best current theories, are the true permanent features which mark the spiral pattern, which is produced by a standing shock wave. The more obvious bright spiral pattern behind the dark lanes is simply a by-product of the presence of this shock. When the Solar System runs into such a dust lane – every 250 million years or so – the immediate result will be a warming of the surface layers of the Sun as it accretes dust. The process is, on a smaller scale, exactly equivalent to the way X-rays are probably generated in binary systems when matter falls into a degenerate star from a companion.

Paradoxically, this warming of the Sun might cause an Ice Epoch to set in on Earth; the idea has previously been investigated by such eminent astronomers as Hoyle and Lyttleton, and McCrea agrees that the effect of a slight increase in solar output on precipitation and cloud cover on Earth could produce a situation in which although more heat is incident on our planet, its increased albedo cause not just the extra heat but a lot more as well to be reflected away.

So far so simple; and the model stands up when McCrea puts in the appropriate numerical calculations for the effect of plausible dust densities on the Sun. Even better, the model explains our present situation in a straightforward manner.

Just now, the Solar System is on the edge of a spiral arm (the Orion Arm)

Fig. 22 Spiral galaxy MSI *(Hale Observatories).*

having emerged from the associated lane of dust and gas about 10,000 years ago. We may even, says McCrea, have passed through the Orion Nebula itself – and the time when we left the dust ties in nicely with the end of the most recent period of intense glaciation, which also ended some ten millennia ago. If McCrea's ideas are correct, it should be all plain sailing around the Galaxy for the next 250 million years!

But the story doesn't end there. McCrea elaborated on his theme in Oxford recently, when he gave the 1975 Halley Lecture. Then, he suggested that the Sun's family of comets is a by-product of the recent passage through a dust lane, produced by compression of the gas and dust of interstellar space in the shock wave. Looking further back in time, perhaps the whole Solar System was created by the repeated passage of a cloud of gas and dust through the shock wave spiral pattern; each passage would produce a further stage of compression, until eventually stars and planets could condense out.

So our Solar System may be the offspring of the spiral arms. These variations are likely to be of great importance to any life forms emerging on a planet, and McCrea suggests that the development of life depends not only on the distance of a planet from its parent star, but on the distance of the star from the centre of its parent galaxy, and the frequency with which it encounters spiral arms. Even for astronomy, a theory whic ties together the origin of the Solar System and comets with the emergence of life on Earth and the occurrence of Ice Ages is imagination on the grand scale.

John Gribbin

Index